THE QUEST FOR COMETS

THE QUEST FOR COMETS

AN EXPLOSIVE TRAIL OF
BEAUTY AND DANGER

DAVID H. LEVY

AVON BOOKS ◆ NEW YORK

AVON BOOKS
A division of
The Hearst Corporation
1350 Avenue of the Americas
New York, New York 10019

Copyright © 1994 by David H. Levy
Published by arrangement with Plenum Publishing Corporation
Library of Congress Catalog Card Number: 94-2741
ISBN: 0-380-72526-6

The Plenum Press edition contains the following Library of Congress Cataloging in
Publication Data:

Levy, David H., 1948-
   The quest for comets : an explosive trail of beauty and danger/David H. Levy.
      p.   cm.
   Includes bibliographical references and index.
   1. Comets—Popular works.   2. Comets—Research—Popular works.   I. Title.
QB721.4.L48 1994   94-2741
523.6—dc20   CIP

First Avon Books Trade Printing: September 1995

AVON TRADEMARK REG. U.S. PAT. OFF. AND IN OTHER COUNTRIES, MARCA REGISTRADA, HECHO
EN U.S.A.

Printed in the U.S.A.

OPM 10 9 8 7 6 5 4 3 2 1

With love and affection, this book is for

My brothers and sister, Richard, Joyce, and Gerry;

My in-laws Larry and Audrey;

My nephews and nieces Jeff, Marci, Wendy, Robby,
Ali, Billy, Debbie, Michael, and Daniel

Preface

I looked forward so much to those weekly dinners with Lonny Baker. At Tucson's Flandrau Planetarium, she was in charge of the lecture series "Eyes on the Universe," which featured a different astronomer each Tuesday evening. Our plan was to meet before each lecture for dinner, and tonight was especially important, for Martin McCarthy, a famous Jesuit astronomer, was talking about "Gregory, Galileo, and Galaxies," a religion and science lecture about creation versus the Big Bang. Nothing could keep me from this lecture.

Nothing, that is, except a clear and moonless night.

For the past 19 years, I had been among the world's thousands of keen amateur astronomers, organizing my life to accommodate my true vocation—searching the night sky for comets, hoping to discover one never spotted before. Although I had devoted 917 hours with my eye at the eyepiece and every dime I could scrape together to further my quest, although I had moved from cloudy Montreal to the Arizona desert southeast of Tucson, although I had observed every known comet I could find, I had yet to find one of my own.

The clouds in the early evening of Tuesday, November 13, 1984, promised Lonny and me a leisurely meal. We met at a Chinese–Mandarin place, and ordered our favorite dishes—no peas, no MSG, Lonny insisted. But as dinner went on, I sensed that the clouds were getting thinner, and Lonny could see that instead of concentrating on her words, I was looking past her out the window. "David," Lonny said, "it must be clearing up outside." "Uh huh," I replied absentmindedly. "You're going to stand me up, aren't you?" she demanded. "You're going to go home and hunt for comets, aren't you?" "Oh no!" I protested, gamely snapping to attention. "We are going to finish dinner. *Then* I am going to stand you up, go home and hunt for comets."

As we left the restaurant, she for her lecture and me for my telescope, Lonny gave me an ultimatum: "You'd better find me a comet tonight!"

Setting up the telescope for comet hunting is pretty easy. All I really have to do is make sure the cover is off. With the telescope's slow, deliberate motion across a portion of sky, comet hunting is not like a star party, where people line up to look at an object. When that happens, the sky is asked to be a servant, showing off Saturn, the moon, or some galaxy on cue. It's the opposite with comet hunting. When I start a session, I have only a vague idea of what I may find in the next hour or so as I move the telescope forward for a few minutes across a region of sky, then backward through the next sector. Whether I find a star cluster or a galaxy, a red star or a bright double star, is really up to the sky, not me. The sky is the master, my telescope the receiver, and I am the watchman.

After some 30 minutes had gone by, a faint fuzzy object appeared in the field of my moving telescope. It had the appearance of a galaxy, I thought, and a quick check of an atlas confirmed my suspicion. Like a fish thrown back into the water, the object was gone, and I was on my way. The next object was a planetary nebula, the remnant of an outburst in an ancient star—interesting, but not my quarry.

Next came a pretty cluster of stars, number 6009 in the *New General Catalogue*, but my attention was drawn to a fuzzy object in the same field of view, a bit to the south. It was a striking sight, the beautiful cluster and the faint fuzzy spot; my first reaction was,

"Why have I never seen such a thing? It should be pictured in all the astronomy books." Another atlas check confirmed my growing suspicion: The cluster belonged there; the faint fuzzy spot did not. Within a few minutes, I was sure that the object was moving very slowly in the direction of the cluster. A comet! My heart rate soared.

It was a comet all right, but was it already known? I called Brian Skiff, an observer at the Lowell Observatory some 300 miles away in Flagstaff. The 5 minutes before his return call seemed like an eternity. By now I was quite agitated. "David," Brian finally said, "You've found a new comet."

I have always wondered what the difference between a successful search program and a failed program would be. That night gave the answer: a single field of view. Over 900 hours at the eyepiece, field after field had failed to show me this holy grail.

I was thrilled beyond words, but now I needed wits more than feelings. I sent a telegram to Brian Marsden, director of the Central Bureau for Astronomical Telegrams (CBAT), the clearinghouse for astronomical discoveries. Included were the comet's position, direction and rate of motion, and brightness. With an independent discovery by Massachusetts amateur Michael Rudenko the following evening, the comet was announced as Comet Levy–Rudenko 1984t, the twentieth comet found or recovered in 1984. It was the first of a series of discoveries from my backyard.

Whether they curve round the sun in orbits ranging from a few years to two centuries or come in just once and then back away, tail first, into the space beyond the planets, comets offer a sense of mystery and romance. They may also hold some interesting secrets about the history of the solar system. But comets cannot be studied if they haven't been found. Each clear night a band of us night watchmen of the sky, spread out in backyards and roadsides all over the world, scans the sky in an intense competition. For us, hunting comets is sport, art, and passion. It is by far the slowest of all sports. It demands time—not time set by the hunter but by the sky itself. The moon acts as a referee, brightening the sky and limiting the period available for searching. The game is especially competitive just after a full moon, when a sky that has been too bright for hunting is suddenly thrust for 1 or 2 hours into darkness, and comet hunters all over the world rummage through it in a scramble for new comets. Two weeks later an almost new moon opens the morning sky for another inning as we sleepy searchers peer through telescopes to see if anything new has appeared.

In summer 1989, having just discovered my fifth comet from my backyard, I joined the asteroid and comet survey being conducted by Eugene and Carolyn Shoemaker at Palomar Mountain. Here was a totally different kind of search. For seven nights each month, I was part of a team photographing the sky through Palomar Mountain's 18-inch Schmidt—a camera telescope. Carolyn scanned films that we took, sometimes as many as 60 a night. By fall 1993, we had found 12 new comets together, 9 of which were periodic, bound to the Sun in short orbits. Combined with my own visual search, I now have 19 comets to my credit. With 30 comets, Carolyn's name is on more comets than anyone else's in history.

Carolyn's husband, Dr. Eugene Shoemaker, is a geologist who studies our neighboring asteroids and comets. Through many cloudy night conversations, I have become impressed with the single-mindedness of his path through life. He ended a controversy and launched the new science of astrogeology by proving that Meteor Crater in Arizona was the result of an impact by an asteroid from space. During the sixties and seventies, he worked with the space program, studying craters on the moon and other solar system bodies—craters, he correctly thought, that came from impacts of asteroids and comets. For his current work at Palomar, Shoemaker has switched from studying the bull's-eyes to the bullets, those objects in space before they hit anything else.

Comet impacts are the substance of this book, but comet hunting is its soul. This 200-year-old tradition is as much competitive sport as science, and its history is filled with drama, strange coincidences, hoaxes, and fun. The story of my own first comet is just one example. After I finished sending my telegram, I remembered that Lonny was still at the planetarium. I called her just as she was about to leave. "Well David, did you find me a comet tonight?" she asked. "Yup." Lonny giggled. "Where?"

I told her it was in Aquila, the constellation of the Eagle, and she laughed again. Then I said that it was moving a minute of arc per hour to the north. Lonny stopped laughing. "You're serious, aren't you?" The next day the local paper told the story on its front page. The piece did not focus on the newly found fuzzy world, whose existence was not deemed newsworthy. What did make news, the paper thought, was that I had broken a dinner date to do it.

Acknowledgments

Few things mean as much to me as comets do. To write about them *clearly and sensitively,* I have tried to merge the personalities, the science, and the events into a single story that includes the poetry of the comet hunt as well as the reality of the story of comets and the major role they played in the history of Earth and its life. Part of the book is autobiography, and for the chapters about Gene Shoemaker and comet impacts, part is biography. And the comets themselves have much to tell. The result is intended to present a feel for what comets are like and what the people are like who work in the comet field.

To accomplish this, I am grateful for all the assistance I received. Linda Greenspan Regan at Plenum was helpful and encouraging throughout the growth of this book. Steve Edberg, Daniel Green, Brian Marsden, and Sandy Sheehy gave invaluable reviews and comments. Gene and Carolyn Shoemaker gave a large amount of their time for interviews and were helpful in many ways. Norman MacKenzie of Queen's University provided important suggestions for the material about Seneca; and L. Bean, Roy Bishop, Clark Chapman, Eleanor Helin, Tim Hunter, Peter Jedicke, Joe Marcus, David Morrison, Jean Mueller, Donna Donovan-O'Meara, David Rabinowitz, Jim Scotti, and Wieslaw Wisniewski were helpful with other parts of the manuscript. Kelly Beatty and Steve O'Meara, from *Sky and Telescope* magazine kindly let me use material from my "Star Trails" column, and they were helpful in many other ways as well. Finally the *Journal of the Royal Astronomical Society of Canada* is publishing my Ruth J. Northcott memorial lecture, "The Art of Comet Hunting," which contains some passages from this book.

☾ ☽

Contents

THE QUEST FOR COMETS

The Terrible Swift Sword
Pleasures and Perils of a Comet

The sky above Quebec's Jarnac Pond was inky, the night silent except for the occasional croak of a frog and the hoot of an owl. I could see more stars than I could count, and the Milky Way arched right overhead. I was 14, and for the past 2 years I'd been increasingly drawn to astronomy. "You are going to sit on the dock all night long? *Until dawn?*" My astonished grandparents couldn't believe my plan. On that lazy afternoon of August 12, we were watching clouds pass by and talking about the very bright meteor we had seen the previous evening. I had read in a book how, each August 12, the Earth's orbit intersects the orbit of a comet called Swift–Tuttle, and over a very long time, dust from that comet had spread all along the orbit, resulting in the Perseid meteor shower.

It seemed so simple. Although the meteors, most no larger than a speck of dust or a pea, would be on parallel paths when they entered the atmosphere, they would appear to radiate from a single point in the sky in the constellation Perseus. With great excitement, I somewhat precociously explained to my grandparents that the sight of meteors coming from a radiant was due to an effect of perspective—just like looking down a railroad and watching the tracks converge. But then came the punch line: As Perseus rose higher, we would see more meteors; the shower would be

Periodic Comet Swift–Tuttle. (Photograph by David Levy using his 8-inch Schmidt telescope from his backyard, November 18, 1991; 5-minute exposure.)

strongest just before dawn. That meant, I insisted, that I was going to stay up all night.

For the rest of my family, staying up late to look at the stars was no surprise. My parents had learned to navigate their sailboat by the stars, and we often spent evenings looking at Saturn through my small telescope. As night fell on August 12, I began my vigil. Grandma and Grandpa sat with me on the dock for a short while, relaxing on deck chairs in the warm evening, but clouds prevented much observing at first. Knowing my reputation for being somewhat klutzy, grandpa had built a makeshift fence on the dock so I wouldn't fall into the lake and drown. As the night went on, the clouds gradually dissipated, and I was absolutely thrilled to see the prediction come true. With each passing hour, the number of meteors increased. One of them appeared to split in two, others left beautiful sparkling trails. Alone on the lake that night, I was treated to a personal show of fireworks in the sky. When dawn finally came, I had logged 112 meteors.

Thirty years of Perseid meteors have passed since that August night, some were truly memorable. In 1966 I was part of a team of observers that spotted 906 meteors. Bright moonlight interfered with some of these observations, and clouds ruined others. In the early 1980s, observers around the world started paying more attention to the meteor shower, for its parent comet, last seen in 1862, was due back. Would the comet's return be accompanied by a large increase in the numbers of meteors?

THE TERRIBLE SWIFT–TUTTLE

When Lewis Swift discovered his first comet on July 15, 1862, he doubted that he had a new one. Instead he thought it was the already known Comet Schmidt, which had passed close to the Earth on July 4 of that year and was rapidly fading. Swift was using a 4-inch refractor telescope from his home in Marathon, New York, near the college town of Ithaca. But three nights later, Horace P. Tuttle saw the same comet from the small balcony attached to the refractor dome at Harvard College Observatory just as he was preparing to join the Union army fighting the Civil War. Finally realizing that his comet was a new one, Swift rushed to try to get credit for it. "I received a letter from Mr. Swift, an amateur astronomer, at Marathon N.S.," wrote G. W. Hough, assistant at Dudley Observatory, "in which he stated he had observed a comet on the 15th and 16th of July; but presuming it to be Comet II.1862 he gave no public notice of the observation."[1] It appears that Thomas Simons also discovered the comet on July 18 in Albany but did not get his name on it. In England it was called Rosa's comet for the observer who first saw it from Rome on July 25.[2]

The 1862 appearance was a memorable one. The comet was as bright as the north star, easily visible to the unaided eye. "The last time I saw the comet," codiscoverer Tuttle wrote, "was on the night of the 17th of September, after I had entered the federal army and gone to camp. It was then about six degrees [about 12 diameters of the full moon] from Antares and just visible to the naked eye."[3]

Just 4 years later, Giovanni Schiaparelli, an astronomer later famous for his observations of Mars, floated the remarkable idea that meteor streams are residue from comets. As an example, he showed that the orbit of the famous Perseid meteor shower was analogous to that of Swift–Tuttle. So not only did Swift and Tuttle have a bright comet to their credit, but they also had the one whose debris would fall in a meteor shower every August.

HORACE TUTTLE'S ORBIT

Once Horace Tuttle's comet became noteworthy, Horace Tuttle became famous, too. But like amateur astronomers throughout time, he had to sandwich in his observing between less lofty duties. Shortly after his comet discovery—it was his fourth—in 1863, he became acting paymaster in the Union Navy.[4] In 1864 Tuttle's ship, the Catskill, was anchored off Charleston, SC to blockade Charleston and Wilmington. From the Catskill's deck, Tuttle saw the comet that had been found by Ernst Tempel as it passed from Auriga into Taurus. As the ship swayed, Tuttle must have found Comet Tempel difficult to watch as it dashed repeatedly through his telescope's field for a few seconds at a time.

Later in 1866 Tuttle shared a comet find with Tempel. Remarkably Periodic Comet Tempel–Tuttle is the mother comet of the November Leonid meteors, the shower that appears to originate in the constellation of Leo the lion. Although Tuttle holds the distinction of having discovered the source comets of two major meteor streams, he wasn't quite so exemplary in other fields.

In 1869 while serving as paymaster aboard the monitor ship *Guard*, he somehow left the ship's books with a mysterious loss of $8800.90—a tremendous discrepancy at the time. Eventually the navy, realizing it had never received the thousands of dollars it claimed Tuttle owed, ordered him to Washington for courtmartial. Stung by a growing embezzlement scandal, the military had just sentenced an army paymaster to life in a penitentiary, commuted to 5 years by President Grant. Tuttle's fear that the same thing might happen to him did not however prevent him from using the Naval Observatory's 26-inch refractor to recover Periodic Comet Encke on January 23, 1875, 3 days into his courtmartial. Three weeks after the comet recovery, the court found Tuttle guilty of embezzlement. Because of doubts regarding his lawyer, and perhaps because of his comet discoveries, Tuttle's sentence was light: a dishonorable discharge from the navy, approved by President Grant.[5]

SWIFT–TUTTLE'S ORBIT

More than 900 comets have passed by since the first recorded comet almost a thousand year before Christ. Most comets come by on wide paths, or orbits, that take them round the sun and then back into deep space, not to return again for thousands of years if ever. But some 200 comets do return in periods of two centuries or less. Although Halley, the most famous of these periodic comets, orbits the sun in about 76 years, other periodic comets have different periods. Until 1992 we were not sure how long the period of Swift–Tuttle was.

Once people who calculate orbits have determined that a new comet is periodic, they predict the time of its next return visit. As that time approaches, astronomers with photographic telescopes, or telescopes with electronic detectors called charge-coupled devices or CCDs, try to locate the comet far out in space.

Computing the orbit of a comet is a little bit like watching an airplane flying above you. Imagine that you are looking low in the sky as the plane speeds along. For the purpose of this analogy, your eyes are closed except for one brief instant when you see the plane over a tree and for a second instant when the plane is over an apartment roof. From these two observations, you will have a poor idea of which direction the plane is headed, for it could have gone from point *A* to point *B* by any number of curved arcs instead of a straight line. However if you open your eyes for a third instant, now seeing the aircraft over a house, you have a far better idea of the plane's path. Ten, thirty, or a hundred positions of the plane at various instants will give you a progressively more accurate idea of the path.

But comet orbits are huge; that of Swift–Tuttle takes it almost 50 times the distance between the Earth and sun. A small uncertainty about the positions of the comet when it nears Earth can result in an error of years in predicting the comet's return more than a century later. The elements of an orbit are explained at the end of this book.

In 1902 William T. Lynn, a British astronomer and inveterate writer of letters to astronomical publications, wrote a one-page memoir about Periodic Comet Swift–Tuttle. In it he made an almost off-the-cuff suggestion: "It is possible that the second comet of 1737 may be identical" with the comet that Swift and Tuttle had discovered in 1862.[6] The 1737 comet was found from Beijing by Father Ignatius Kegler, a Jesuit priest. Since Kegler observed the comet for hardly more than a week, we have only a rough idea of its orbit, and Lynn evidently considered his estimation to be hardly more than a conjecture. But we would not know if the two were the same comet until it went to the depths of the solar system beyond the planet Neptune and then returned once more.

When Periodic Comet Swift–Tuttle finally did return, astronomers tried to calculate the details of its next pass and to their amazement, one of the calculations showed that on July 11, 2126, this comet might come much closer to Earth than ever before—perhaps disastrously close. In fact for a few months, some astronomers worried that this large comet would hit the Earth in the largest explosion in 65 million years, destroying most life forms. What comes next is the story of the process leading to that brief period of fright.

Using the positions of Periodic Comet Swift–Tuttle at many instants during its flyby in 1862, in 1973 Brian G. Marsden, director of the International Astronomical Union's CBAT, projected that the comet would loop beyond the orbit of Pluto and return between 1979 and 1983, most likely in 1982. However he was troubled by the last observations of the 1862 appearance, made from the Cape Observatory in South Africa, an observatory in the southern hemisphere that can see a large region of sky never visible from northern observatories. These observations did not fit the path that the comet had earlier appeared to be taking. It was almost as if the airplane's "pilot" had deliberately changed direction—not by much but enough to threaten the accuracy of the orbit Marsden had computed.

What if the comet did have a pilot capable of making such changes? Not a real pilot, of course, but, say, that as the comet rounded the sun, its surface started erupting, with jets of gas or dust spewing forth like steering rockets on a spacecraft. In the science of celestial mechanics, these eruptions are called nongravitational forces, since they can cause lane changes that have nothing to do with the gravity of the sun or the planets that traditionally define the comet's orbit. These forces can be strong enough to push the comet around slightly, changing its course in ways that are very difficult to calculate.

If these forces are too strong, the comet's orbit cannot really be pinned down without observations from at least two passages around the sun. Thus in 1973, Marsden spent several weeks looking at orbits of other eighteenth-century comets that could have been in fact early appearances of Swift–Tuttle. (Later we see that Edmond Halley did much the same thing to show that some comets were actually early appearances of the comet of 1682, the comet that would later be named Halley.) The first possibility was Wargentin's comet in 1750. Although it was within a few degrees from where Swift–Tuttle might have been on that previous pass had the 1862 orbit been correct, Marsden noted that this comet was moving ten times too rapidly. Marsden then suggested that Kegler's comet could be the one. Trying to connect the 217 observations of 1862 with Kegler's week-long view in 1737, Marsden came up with a prediction that the comet could return to perihelion—its closest point to the sun—around November 25, 1992. But instead of the 120-year period suggested by the 1862 appearance, if the comet was indeed Kegler's, the period must be more than 10 years longer.[7] That discrepancy alone prompted Marsden to doubt the connection. However if he were right—if Comets Kegler and Swift–Tuttle were the same—then it could not possibly return in 1981.[8]

It didn't. However most astronomers doubted the connection Marsden had tried to make, assuming instead that the comet had somehow slipped by without being detected.

SWIFT–TUTTLE RETURNS

On August 12, 1991, my friend Peter Jedicke and I were observing Perseids from a mountaintop in southern Vermont. Unlike the beautiful lakeside night almost 30 years earlier, this night was mostly cloudy. However we were impressed by the number of meteors we could see through holes in the clouds. One really bright meteor raced across the sky, its glow lighting up the cloud from atop like a searchlight beam. But that was nothing compared with what happened a few hours later on the other side of the world:

A group of observers near Japan's Kiso Observatory got the impressive rate of 352 meteors in a single hour of observing.[9] The next day, I happened to be visiting Marsden in his office at the Harvard–Smithsonian Center for Astrophysics. I had rarely seen him so excited about an observation. Looking like a kid on Christmas morning, he wondered if it were possible that his 18-year-old prediction just might come true. Was this hail of meteors spotted from Japan a vanguard? Was Swift–Tuttle on its way at last?

Despite a full moon that washed out the sky during the week of the Perseid shower in August 1992, observers tried to see what activity there was. On a beautiful clear night in Montreal, I spotted a paltry two meteors, but then another burst emerged—this time at precisely the moment the Earth crossed the plane of the comet's orbit. Despite these hints, Marsden was now starting to become shy about his prediction. Would the comet appear at all? "September maybe, but . . . nah! it won't come!" Marsden confided, "I got cold feet."

Nevertheless Marsden admitted later, "Another computation I made on July 25 gave T (or time of perihelion) = 1992 Dec. 11."[10] Later he added, "Perhaps it was that, with the exception of Kohoutek, no professional astronomer seemed to share my faith in it . . . or perhaps it was just that the day of reckoning was almost here."[11]

The answer came at last on September 26. Observing from Nagano, Japan, Tsuruhiko Kiuchi, an amateur astronomer using 5-inch diameter binoculars, spotted the comet on a path that would return it to perihelion only 17 days later than the date Marsden had predicted almost two decades before, and only 1 day off his July calculation.[12] By the following day, Jeremy Tatum from British Columbia had confirmed the sighting and had provided accurate measurements to show that this definitely was Periodic Comet Swift–Tuttle. "Things can be as exciting on paper or computer screen as at the eyepiece," Marsden said. "My whole prediction for P/Swift–Tuttle [P/ is an abbreviation for periodic comet] back in 1973 was done by paper and computer (before they had screens!), although it did take someone at the eyepiece to get the final proof."[13] Comets Kegler and Swift–Tuttle were indeed one and the same.

SWIFT–TUTTLE'S FUTURE

With observations now spanning three centuries, Marsden set out to determine whether other past appearances, like the comet of 69 B.C., might have been Swift–Tuttle—and also to predict what the next return to perihelion in 2126 would be like. "Whenever perihelion passage occurred between late June and early September," Marsden had written in his now-famous 1973 paper, "the comet should have become at least as bright as in 1862, and one would expect this to happen during one revolution out of every five or six; near collision with the Earth would take place if the comet were at perihelion in late July."[14]

When Marsden calculated the orbit forward to 2126, he came up with a perihelion date of July 11. Were the Earth and Swift–Tuttle on a collision course? What about those nongravitational forces? These could easily cause a delay. It did not take much computer time for Marsden to determine what might happen if the comet reached perihelion during a particular 3-minute period on July 26. A couple of weeks after perihelion, the mighty comet's nucleus, as large as 10 kilometers across, could crash into the Earth on August 14. Shortly after Marsden made that calculation, I wrote to congratulate him on being written up in *Time* magazine for his successful prediction. "Yes," an excited Marsden wrote back, "we've been getting a bit of *Time* exposure—and we may get more when we remark on the distinct possibility that P/S–T will become the 'mother of all Perseids' and hit us when it returns in 2126!"[15]

Even though the chance for a collision was almost nonexistent, reporters inundated Marsden with questions. "Life has been one big press conference ever since the middle of August," Marsden noted, hoping that the publicity would inspire astronomers to continue to make careful observations of the comet as it moved away from the sun.[16]

With the casual remark on Circular 5636 in October 1992 that Periodic Comet Swift–Tuttle may hit the Earth on its next return, quite a few astronomers have wondered if Marsden had perhaps gone a bit far by suggesting that there were a one in 10,000 chance of a collision. Not so, he says:

> My message was to astronomers and the need for them to observe the comet during the next several years. The observations in 1862 showed that Swift–Tuttle behaved in a very peculiar fashion—something of the kind I have never seen before in nearly 40 years of computing orbits. I regret the probability figure; that was not on the IAU circular and was rather forced out of me by reporters. The fact is that, even if Swift–Tuttle doesn't get us next time, it will have ample opportunity to do so in the more distant future, and it is the largest object we know of that has the potential to strike the Earth.[17]

By the end of 1992, Marsden reported that comets that appeared in 69 B.C. and A.D. 188 were actually early appearances of Swift–Tuttle. With the comet's orbital history now going back more than 2000 years, Marsden could now state with great certainty that the comet would return to perihelion pretty close to July 11, 2126, and there was now no chance of a collision with Earth. The scare was over—at least until 3044, when an even closer approach is predicted. Unless some new comet with our name on it rounds the sun in some surprise attack, Periodic Comet Swift–Tuttle is the largest comet on an orbit known to intersect the Earth's orbit. Thus there is some chance—albeit a remote one—of a comet hitting the Earth sometime within the next several hundred thousand years. Although the comet is only 15 kilometers in diameter, the speed of its impact would create a crater at least 200 kilometers wide, destroy the ozone layer, send up a pall of dust that would block sunlight for months, and destroy most species of life on Earth.

When the comet returns in 2126, we may get a special show anyway—from both the comet and its associated stream of Perseid meteors. Instead of its usual hundred or so meteors per hour for a few hours that night, the sky should blaze with a spectacular fireworks as thousands of meteors hit the atmosphere. Perhaps someone will then remember how, back in 1992, we thought for a while that it might be all over for us.

WHAT ARE COMETS MADE OF?

An ordinary comet has three parts—a nucleus, a coma, and often a tail. The nucleus is the source of everything we see in a comet—the dust and gases that form the coma and tail. Although observers have detected a few comet nuclei, our most dramatic view came in March 1986 when three camera-laden spacecraft—the European Space Agency's *Giotto* and two Soviet *Vegas*—sped by Halley's comet as it crossed the plane of the Earth's orbit on its way out of the inner solar system. The spacecraft recorded what looked like a large potato some 8 kilometers wide and 15 kilometers long—a really small object to generate a coma hundreds of thousands of kilometers across and a tail several million kilometers long. As some of the solar system's tiniest members, comet nuclei produce comas and tails that are among its largest and most beautiful features.

The three Halley spacecraft also confirmed what Harvard's Fred Whipple had proposed in 1950: A comet nucleus is like a snowball with dust mixed in. Instead of being a dirty snowball however, Halley's comet contains so much dust that it is better called, as astronomer Mark Sykes suggests, an icy mudball. The snow is not pure water ice but a mixture of frozen gases, including carbon monoxide (CO), hydrogen cyanide (HCN), and other ices containing carbon and sulfur.

When a comet is far from the sun, the nucleus is all there is. But as it nears the sun, the comet's ices—from water and other materials—begin to change into gas in a process called sublimation. We see a comet in two ways: Its dust particles scatter light from the sun, and ultraviolet light ionizes the gases and causes them to fluoresce. Comet Halley's nucleus loses material at a rate per orbital revolution that if spread all over the nucleus, would be 1 meter thick. In fact even when close to the sun, most of the nucleus is quiet except for active areas that are the primary source of material in the coma.

The coma is the atmosphere of material that surrounds the comet's nucleus. Gas streams from active areas at high speeds—several hundred meters per second—and carries dust particles with it into the tenuous cometary atmosphere. Most comets form comae when they get as close to the sun as the asteroid belt, well beyond the orbit of Mars. As Periodic Comet Shoemaker–Levy 9 has so dramatically shown after its encounter with Jupiter in 1992, a comet can have more than one nucleus and coma.

The coma and nucleus together make up a comet's head; material that streams away from the head is called the tail. The sublimated gases form a gas or ion tail, which generally points nearly in the direction opposite the sun, and the dust follows the comet at a more leisurely pace, often curving away slightly. As the dusty material blows away, the comet orbits the sun more slowly and forms a curved tail as the more distant particles lag further behind.

Comets are surrounded by large hydrogen clouds, as was found in January 1970 when Comet Tago–Sato–Kosaka became the first to be studied by a spacecraft. The craft discovered a huge cloud of atomic hydrogen gas emitting ultraviolet light. Depending on the size and activity of the comet and its distance from the sun, the hydrogen envelope can be 1–10 million kilometers long.

When Beggars Die

Maybe the Roman empress Calpurnia was right, comets can indeed bring calamity—but as we now know, only if they actually strike the Earth. For most of human history however, comets portended major disasters, such as the demise of a head of state, as Calpurnia reminded her husband in Shakespeare's *Julius Caesar*:

> When beggars die there are no comets seen,
> The heavens themselves blaze forth the death of princes.[1]

BROOM STARS AND BUSHY STARS

We are fortunate that comets in ancient times received such attention, for whenever one appeared, both its track across the heavens and its physical appearance were recorded. During the war between two Chinese kings, Wu-Wang and Chou, around 1059 B.C., a comet with an eastward-pointing tail dominated the morning sky. The scribes' detailed account is the first known record of a comet. Chinese recorders eventually noted two types of comet, the *po* and the *hui*. The *po*, or bushy star comet, generally meant a comet with a large fuzzy coma or atmosphere, usually without a tail, the *hui*, or broom star comet, had a tail. The Greek philosopher Aristotle called them fringed and bearded stars, respectively.

In recent years, we have had a cacophony of bushy star comets, one of which I was fortunate enough to discover in May 1990. At first Comet Levy 1990c moved almost imperceptibly among the stars of Pegasus, then in the northeast morning sky, as it closed in on Earth and sun. But as Comet Levy got closer, it brightened rapidly and picked up speed. By the middle of August, Comet Levy was parading elegantly across the summer sky, looking like a fat fur ball as it moved alongside the Milky Way. Although Comet Levy had a tail, it was not easily visible except in photographs. We couldn't see the tail at its best because it pointed away from the Earth.

The best recent example of a broom star comet was Comet West, which punctuated the morning sky like a huge exclamation mark in March 1976. I will never forget the frigid morning that my friend Carl Jorgensen and I drove south of Montreal's city lights to see what we thought might be a fairly bright comet. As we approached our observing site, I looked toward the east, but our view was blocked by trees. As soon as there was a break in the roadside forest, I looked eastward and saw the comet with its long tail pointing upward. "My God!" I exclaimed, "it's magnificent!" Still properly keeping his eyes on the road ahead, Carl agreed, sort of. "We'll get a good view," he promised. "We'll get to the site and see it through the binoculars." "No," I cried, barely able to keep from leaping out of the moving car. "It's right there!" "We'll see it soon," Carl said. "O.K." Carl drove another hundred yards and then casually looked to his left. "My God!" he finally agreed, and almost drove the car off the road.

What if I had been an astronomer in ancient Rome and this magnificent stranger had appeared in the familiar night sky? Even knowing it was a ball of ice and dust flung by gravitational forces into its orbit around the sun, I was almost incoherent with awe. Like most modern comet observers, Carl and I were elated to see this broom star comet. It was far brighter than we had expected. We sketched it and photographed it, remaining at our roadside site until the brightening dawn drowned out its light. Seeing this ghost from space was an emotional high, but although the presidency was being fought over in the United States that year, no one suggested that this spectacular comet had been sent to announce that President Ford would lose. Nor did anyone, as far as I know, blame Halley's comet in 1986 on the fall of Baby Doc Duvalier in Haiti or of Ferdinand Marcos in the Philippines.

THE HEAVENS BLAZE FORTH THE DEATH OF PRINCES

Over the thousands of years of recorded history, comet study has been free of superstition for only the last two centuries. The first book of *Chronicles* describes what could be a comet—the comet of 971 B.C. appeared around that time—that protested an ill-advised census King David had ordered. The biblical passage is read every year at the Passover seder: "And David lifted up his eyes, and saw the angel of the Lord standing between the earth and the heaven, having a drawn sword in his hand stretched out over Jerusalem."[2]

Caesar's wife Calpurnia may have seen a comet, but not *before* her husband Julius was murdered on the Ides of March 44 B.C. at the foot of Pompey's statue. Two months later, a bright comet with a tail perhaps 12 degrees long—half the length of the Big Dipper—was reported moving north to south. This comet is one of the best known from ancient times, but not because it was exceptionally bright or long-lasting. According to Plutarch, "Among the divine portents there was also the great comet; it appeared very bright for seven nights after the murder of Caesar, then disappeared." Calpurnius Siculus went further, blaming the comet for the civil war that followed: "When, on the murder of Caesar, a comet pronounced fatal war for the wretched people."[3]

In Roman Times, comets were known for what they meant rather than for what they are. Rather than objects to be admired or studied, comets were seen as messages, and in the social disruption following Caesar's death, people would see a comet playing a big role. In a beautiful treatise called *De Cometis*, a part of his

Quaestiones Naturales, Lucius Annaeus Seneca also wrote of the comet of 44 B.C. When it came to astronomy, Seneca had few equals in the ancient world, but he was no match for the treacherous politics of his era. Born around 4 B.C., Seneca lived during some of Rome's worst times, during the reigns of emperors as cruel and incompetent as Caligula, Claudius, and Nero. Seneca was Nero's tutor, and when Nero, at age 17, became emperor of Rome, Seneca had considerable power in the government of the state. For 10 years the arrangement worked, then Nero murdered his mother Agrippina in 59 A.D. and coerced Seneca into excusing this heinous act. Seneca spent the last years of his life with his friend Burrus trying to limit the excesses of Nero's madness. In 63 A.D. Burrus died, and Seneca tried to resign from politics. Feigning that he still respected Seneca, Nero refused his resignation.

Although Seneca then abandoned his wealth to live in poverty, he knew his end under Nero's tyranny was near. He desperately tried to praise Nero wherever he could: "There is no reason to suppose," he intoned,

> that the recent one [comet] which appeared during the reign
> of Nero Caesar—which has redeemed comets from their bad
> character—was similar to the one that burst out after the death
> of the late Emperor Julius Caesar, about sunset on the day of
> the games to Venus Genetrix.[4]

The soothing words didn't work. In the year 65, Nero accused Seneca of participating in a plot against him and ordered the scholar to prepare for death. According to custom, this gave Seneca his choice of demise. He chose to cut his wrist and bleed to death. Apparently his luck ran out after the comet appeared.

Still Seneca had at least completed *Quaestiones Naturales*, his seven-book opus on natural philosophy, some 5 years earlier while his position was in turmoil and his life in jeopardy. The last section is called *De Cometis*. "No man is so utterly dull and obtuse," Seneca began, "with head so bent on earth, as never to lift himself up and rise with all his soul to the contemplation of the starry heavens, especially when some fresh wonder shows a beacon-light in the sky."[5] *Quaestiones Naturales* was lost for more than a thousand years; its discovery in the twelfth century allowed scholars to understand the ideas of other Greek and Roman scholars about comets. Seneca's insight into comets is remarkable for its time.

> Blind to all the celestial bodies, each asks about the newcomer; one is not quite sure whether to admire or to fear it. Persons there are who seek to inspire terror by forecasting its grave import. But by my honour, no one could embark on a more exalted study. . . .[6]

Seneca perceived comets as an object of study rather than a superstition. Seneca follows with a summary of cometary thought up to his time. He noted that before Posidonius, who was a Greek scholar from Byzantium, Epigenes identified one type of comet "with hair on all sides" (the bushy star of the Chinese); the other "extends a loose kind of fire in one direction" (the Chinese broom stars). Unfortunately little is known about Epigenes. Even when he lived, possibly in the fourth century B.C., is uncertain. Epigenes insisted that the comets "with hair on all sides" were stationary among the stars. Either Epigenes or Seneca may be confusing these with a different kind of sky surprise called a nova. From time to time, a star may undergo an outburst, becoming very much brighter. These outbursts, called novae, are not really new stars but old ones undergoing brief increases in brightness.

One of the liveliest aspects of Seneca's writing is the vigorous way that he refutes those who disagree with him. "It requires no great effort," he sniffs, "to strip Ephorus [who lived around 340 B.C.] of his authority; he is a mere chronicler." Seneca accuses Ephorus of careless reporting: The Greek astronomer "asserts that the great comet [possibly the one of 373 B.C.] which, by its rising, sank Helice and Buris, which was carefully watched by the eyes of the whole world since it drew issues of great moment in its train, split up into two stars; but nobody besides him has recorded it."[7] Perhaps this is so, but other comets have been widely observed to split, including Periodic Comet Biela in 1846; Comet West in 1976, which broke into four parts; and Comets Levy and

Shoemaker–Holt, seen in 1988, and Periodic Comet Shoemaker–Levy 9 in 1993. For an earlier example, Byzantine records show that around 822 A.D., "A comet was seen in the sky as a sort of two moons joined together brightly, and moreover separated by different attachments."[8] Comet splitting is not uncommon.

Seneca believed that comets are formed "by very dense air, and since the most sluggish air is in the north, they appear in greatest number" in that direction. "For the recent one which we saw during this joyous reign of Nero [Seneca was desperate to insist that the summer comet of 60 A.D. was not a bad omen for Nero] displayed itself to view for six months, revolving in the opposite direction to the former one [the comet of 54 A.D.] in Claudius' time." Seneca now describes rough paths for the two comets, that of 54 A.D. "rising from the North up toward the zenith made for the east, always growing dimmer." On the other hand, the comet of 60 A.D. "began in the same quarter, but making toward the west, turned finally towards the south, where it withdrew from view. No doubt the former found moister elements, more suitable for its fire, and pursued them; the latter in turn chose a richer and more substantial district."[9]

Although comets are distributed across the sky almost at random, in Seneca's own experience, comets coincidentally did favor the sky in the north. He undoubtedly observed the comet of June 54 A.D., blamed for the death of Nero's predecessor Claudius, as it passed through Gemini, which is in the northern sky. Another comet the following year also appeared in the sky north of Cancer, and one in 60 A.D. may have passed near the north celestial pole.

Seneca echoed his favorite Stoic philosophers in his view about the nature of comets; however in an important way, he went beyond the traditional Stoic supposition that comets were temporary events. "I rank it among Nature's permanent creations," he declared.

> In none of the ordinary fires in the sky is the route curved; it is distinctive of a star [meaning a planet] that it describes a curve in its orbit. Whether other comets had this circular orbit I cannot say. The two in our own age [the comets of 54 and

60 A.D.] at any rate had. . . . A comet has its own settled position. For that reason it is not expelled in haste, but steadily traverses its course; it is not snuffed out, but takes its departure.[10]

In writing about the fourth-century B.C. scholar Apollonius of Myndos, Seneca stumbled on more of the truth, although the cometary conventional wisdom of the time prevented him from recognizing it. A comet, Apollonius thought, is "a distinctive heavenly body, just as the sun or the moon is."[11] He even explained how comets brighten as they approach the Earth, then fade as they depart. Seneca tried boldly to refute this part of Apollonius' argument. If this is the case, he wondered, then why are some comets at their brightest when they first appear on the scene? Considering the era's scant understanding of the comet's orbits, Seneca's reply is a good one. With a telescope and the ability to compute orbits, Apollonius could have answered that some comets may approach from behind the sun, brightening as they arrive, but remaining entirely unseen until they suddenly appear at their maximum brightness. Seneca's account of Apollonius is the only surviving record of this Greek philosopher's precocious insight.

Years later the historian Tacitus looked back on the days of paranoic Nero and his reaction to the comet of the summer of 60 A.D.

A phenomenon which, according to the persuasion of the vulgar, portended change to kingdoms: hence, as if Nero had been already deposed, it became a topic of inquiry, who should be chosen to succeed him.[12]

Even though the comet predated his book, Seneca was still desperately trying to stay on Nero's good side, even though the people were whispering that Rubellius Plautus, a relative of Julius Caesar, would replace Nero as emperor. Unfortunately, Tacitus notes, the comet's appearance was punctuated by another sign that hit close to home: "As Nero sat at meat in a villa called Sublaqueum, upon the banks of the Simbruine lakes, the viands were struck by lightning and the table overthrown. . . ."[13]

The famous appearance of Halley's comet at the height of the Norman conquest of England in 1066 underscored the perception that the heavens themselves still blazed forth the death of princes. "Nova stella, novus rex!" ("New star, new king!") was the battle cry on the Bayeux tapestry that later depicted the battle. By the fifteenth century, that maxim had still not changed. In a story from the *Illustrated London News*, there is a documented case, in fact, of a comet actually causing the death of a prince. The guilty comet appeared in 1402 and was visible in broad daylight for a record duration of 7 days.

> There is no doubt, however, that comets sometimes really did produce fatal effects. In June, 1402, one appeared in Italy which literally killed the famous John Galeas Visconti. The astrologers of the Prince had predicted that his death would be announced by a comet of extraordinary magnitude, and the celestial phenomenon had no sooner become visible than his Highness, speechless from fright, sank to the ground and died.[11]

Thinking that the comet foretold his death, poor Visconti probably had a heart attack when he saw it. Small wonder. Even with my own modern perspective on the nature of these solar wanderers, I felt my own heart jump and race that night in 1976 when I saw Comet West in all its fiery array arch over the dark Quebec woods.

❦ 3 ❧

Taming the Shrew

On the night in 1864 that Horace Tuttle was observing Tempel's comet from aboard the swaying deck of the Catskill, *a young* English poet was watching the same comet from another shore. Gerard Manley Hopkins later became one of the nineteenth century's best known English poets—his sonnet "The Windhover" is widely read in many schools. The moonless predawn sky was clear on both the English and U.S. coasts; although Jupiter and Saturn had already set, Mars was prominent in the southern sky. High in the east were Auriga and Taurus. Just west of the second magnitude star Beta Tauri shone the head of Tempel's comet, its tail stretching toward nearby Iota Aurigae.

Less than two weeks later, Hopkins wrote these lines:

—I am like a slip of comet,
Scarce worth discovery, in some corner seen
Bridging the slender difference of two stars,
Come out of space, or suddenly engender'd
By heady elements, for no man knows:
But when she sights the sun she grows and sizes
And spins her skirts out, while her central star
Shakes its cocooning mists; and so she comes
To fields of light; millions of travelling rays

Pierce her; she hangs upon the flame-cased sun,
And sucks the light as full as Gideon's fleece:
But then her tether calls her; she falls off,
And as she dwindles sheds her smock of gold
Amidst the sistering planets, till she comes
To single Saturn, last and solitary;
And then goes out into the cavernous dark.
So I go out: my little sweet is done:
I have drawn heat from this contagious sun:
To not ungentle death now forth I run.[1]

As both a poet and an observer, Hopkins wrote about comets in human terms. The comet first appears "bridging the slender difference of two stars." While doing my 1979 Master's thesis at Queen's University, I found a letter to the London *Times* in August 1, 1864, that predicted: "On Monday night it will be situated about five degrees to the left of the Pleiades, passing thence between the stars Iota in Auriga and Beta in Taurus . . ."[2]

On first reading, Hopkins was asking some dated questions: Do comets come from space or are they born in our atmosphere? And what about the line that says Saturn is the most distant planet? As it turns out, neither are out of place in this poem, because Hopkins had intended his poem to be part of a play set in Italy during the Renaissance. His poem was supposed to reflect the uncertainty about the nature of comets that prevailed at that earlier time.

It was Aristotle's theory that comets were, as the poem so appropriately notes, "suddenly engender'd by heady elements" high in the atmosphere. For almost 2000 years, Aristotle's *Meteorologica* was the primary reference whenever people discussed comets. Aristotle believed that comets form when the Earth breathes hot, dry air into the upper atmosphere. It is a view that Seneca tried to reckon with—once the exhalations were formed, he thought, they would be permanent. However, Aristotle's view lasted for so long that it became very difficult to challenge, and once adopted by the Roman Catholic Church as dogma, challenging Aristotle's views could be fatal: For his unorthodox beliefs about the existence of other worlds, Giordano Bruno was put to death at the stake in 1600.

The sudden appearance of a supernova in 1572 knocked a hole in one of Aristotle's main ideas; the heavens did not permit change. Just 5 years later, a bright comet appeared, and 3 years after that, another comet provided gist for new studies on the nature of the heavens.

Hopkins's line about the comet shaking "its cocooning mists" probably recalls a different comet, a large one that appeared suddenly over England in the early summer of 1861. As its nucleus rotated in space, its more active side turned toward the sun at set intervals, resulting in periodic eruptions of material—the "cocooning mists" in the poem. These eruptions were seen by many observers, including William Ellis, an observer assigned to a telescope at Greenwich Observatory. Ellis observed this comet in secret, since his boss, Director George Airy, was a stickler for planning and details and objected to any break in the observing routine, even one caused by a huge comet rising unexpectedly over the horizon. In 1986 Halley's comet shredded its cocooning mists so regularly that they were successfully predicted before the spacecraft visited the comet. Hopkins would have been pleased at the importance of his cocooning mists.

THE COMET PROCESSION OF 1618–1619

A procession of bright comets arrived starting in late summer of 1618. The first, which we now call 1618 I, was found by observers in Kaschau, Hungary, on August 25, as well as by Johannes Kepler in Austria. It was a splendid morning object in the constellation of Leo with a long tail pointing to the west. "In the month of August," wrote Jacobus Mascardus, a Jesuit priest, "news was brought to us from many parts of Italy that during that period a comet was seen licking the hind feet of the Great Bear."[3] Kepler took advantage of his new telescope to observe this comet, probably the first telescopic observation of a comet. By September 15 the tail had all but disappeared, and the rest of the comet was fading.

By the middle of November, a second comet highlighted the southwestern sky; its head was in Libra, and its tail stretched for 40 degrees in length (80 full-moon diameters!) A third comet emerged at the same time, creeping across the sky from Libra toward the west. The final act began on February 14, 1619, with the appearance of a fourth comet in the southeastern morning sky with a tail covering more than half the sky. The comets led to a debate on the nature of these objects between Galileo and the Jesuit astronomer Horatio Grassi. Comets have circular orbits around the sun, insisted Grassi, trying to fit comets into the new system proposed by Tycho, where the planets revolve around the sun, but the sun in turn revolves around the Earth. Galileo, who had maintained a publishing silence since his stunning telescopic observations in 1609, so disputed these ideas that he came out of retirement and resumed writing.

In 1623 Galileo published *Il Saggiatore* (*The Assayer*) in the form of a letter. From today's viewpoint, it is quite a disappointment. Considering that only 9 years earlier its author had shaken the world with his observations of Jupiter's moons, Venus's phases, and sunspots, it seems incredible that *The Assayer* doesn't mention that he observed any of these comets through his telescope. During the fall of 1618 however, Galileo was bedridden with arthritis and a hernia and quite unable to use his telescope.

In *The Assayer* Galileo disagreed fiercely with Grassi's views. Comets could not be in circular orbits, he sneered, since it is clear that all comets do not return again and again. Instead Galileo suggested that comets were formed by sunlight reflected from thin vapors. This would seem to put Galileo on the wrong side of history; however Galileo's modern translator, Stillman Drake, argues that in a sense, he was right, for comets do not shine by their own light. It was Tycho's system of regular orbits, not the nature of comets, that Galileo was fighting.

So bitterly did Galileo dispute the views of his Jesuit colleagues that the rift never healed. But the old scholar's volume was a best seller. As a result, Galileo decided to proceed with his *Dialogue of the Two World Systems*, the work that led to his battle with the Inquisition that resulted in his arrest, trial, imprisonment, and a verdict that was not rescinded until 1992, more than 350 years later.

No matter whether their arguments were right or wrong, their allusions were beautiful. Mario Guiducci's *Discourse on the Comets* describes comets behaving "like Penelope unraveling the cloth with one hand as fast as she weaves it with another."[4] Centuries later Hopkins would put a similar allusion into his comet poem, as his brightening comet "sucks the light as full as Gideon's fleece."

Of all the writings of the comet controversy of 1618, I found that the close of Grassi's description of the comets to be the most touching. The long saga of human relations with comets had reached a new level of understanding: "I have believed that the comet," he wrote, "shining on all directly from the same place and appearing the same from all sides, must be considered as worthy of the heavens and very near to the stars." With a flourish Grassi concludes with a quote from Horace: "With my head exalted I shall touch the stars."[5]

By the early seventeenth century, comets were understood to be objects far away from Earth's atmosphere. One writer stated in 1635,

> Neither was it this starre alone, but others also after it, even Comets themselves, whose places were found to be above the moon: for observing more diligently and exactly than in former times, the observers could easily demonstrate this truth also. . . . If Comets be burnt, consumed, and wasted in the starrie heavens, it seemeth that there is no great difference between them and things here below.[6]

EDMOND HALLEY AND THE COMET OF 1682

As I write Chapter 3 at the end of 1992, Halley's comet is almost as far from the sun as Uranus. Its quiet potato-shaped nucleus, some 15 × 8 × 8 kilometers across, is no longer surrounded by the magnificent coma and tail it brandished like a peacock in 1986. Halley's comet will move even further out, lurking in the darkness of the outer solar system well beyond the orbit of Neptune until about 2055 when it crosses the orbit of Uranus on its way to its thirty-ninth observed return.

With its once-in-a-lifetime schedule, Halley's comet brings generations together. The scene of grandparent and grandchild standing together to look at Halley's comet is repeated every 76 years as the comet loops around the sun, checking in a sense on our progress. As comet hunter Leslie Peltier noted, the comet presided over the defeat of Attila the Hun in 451 and terrified the warriors of the Norman conquest of England in 1066. In the year 1456, Peltier's soaring prose goes on,

> The menacing appearance of the comet so alarmed Pope Ca-lixtus that he decreed several days of prayer and established the midday angelus. With a great clanging of bells he then besought the comet to visit its wrath solely on the invading Turks. In 1607 it was watched by both Shakespeare and Kepler and I like to think that it was also seen by Captain John Smith and Pocahontas in the frontier skies of Jamestown. On its following trip around in 1682 the comet was observed by Halley himself, who probed into its periodic past and bequeathed to it an honored name that it can bear with pride throughout the solar system.[7]

In 1835 the comet presided over the trailblazers heading west to expand the young United States, and in 1910 the comet danced by to a comet rag, and comet pills taken to ward off its supposed effects. In 1986 we met it as never before. The International Halley Watch coordinated many thousands of observations as the comet passed, and the Soviet Union, Europe, and Japan sent a flotilla of spacecrafts to meet it. I started a venture called "Project 2061" in which teachers and children were given directions to find the comet and instructions to remember their traveling friend until it returns in 2061.

Finally just 8 years after Periodic Comet Swift–Tuttle arrives, Halley's comet will roar by in 2134 in its closest visit in 1300 years. But Peltier asks what wonders of technology we will have to study the comet then "Or will man himself prove periodic?" "Will the Huns be back again?"[8]

When the breakthrough finally came that allowed us to see comets as full-fledged members of the solar system, it had nothing to do with the appearance or structure of comets but with their orbits. After three major comets appeared in 1680, 1682, and 1683, Edmond Halley thought to calculate their orbits based on accurate observations made by the astronomer John Flamsteed. Halley also determined the paths of 21 other comets that appeared between 1337 and 1698. His results, which he reviewed with his friend Isaac Newton, were astonishing. The comets that appeared in 1531, 1607, and 1682 had such similar orbits that Halley believed they were actually returns of the same comet. At first Halley suspected that other pairs of similar orbits were returns of other comets, but those never panned out. Halley capped his work with a prediction that his comet would return at the end of 1758. Unfortunately he didn't live long enough to see his forecast come true. During 1758 a mélange of mathematicians and astronomers were frantically searching for the comet on paper and in the sky, but it was not until Christmas night 1758 that Johann Georg Palitzsch, a farmer and amateur astronomer, searching with a small telescope near Dresden, first spotted the comet. With his unaided eye that frigid night, he was pretty sure he saw a faint fuzzy spot low in the sky in Pisces. With some agitation he tried to set up his long telescope, but the cold and his nerves cost him precious minutes. Finally his telescope was ready, and he aimed it at the new object in time to draw it position relative to some nearby stars in his field of view.

Halley did have his work cut out for him 2 centuries ago. "He really had the benefit only of the few cometary orbits he computed,"[9] notes Marsden. In addition to the three returns Halley worked on, Marsden adds that Halley suspected that the comets of 1378 and 1456 were earlier appearances of the same comet.

Periodic Comet Halley, 1986 III. (Photograph by Jack Newton, January 11, 1986, from Victoria, B.C., Canada.)

Periodic Comet Halley, 1986 III. (Photograph by Jack Newton, with a 300-mm f/8 lens, April 1986, as the comet moved through the constellation of Scorpius.)

Periodic Comet Halley, 1986 III. (Photograph by Jack Newton, April 14, 1986, from Cuzco, Peru.) Note that the tail has become much wider as the Earth passes through the plane of the comet's orbit.

It was far more difficult to link Halley's comet with comets in the more distant past. Although a bright comet appeared in 1301, Halley thought that the comet he was studying should have returned 4 years later. But it turned out that the comet's orbital period was varied from 74–79 years, depending on perturbations from the planets. It was not until the nineteenth century that celestial mechanicians like J. Russell Hind were able to connect earlier apparitions to Halley's comet, and early this century, Cowell and Crommelin used ancient records to confirm the comet's visit as far back as 240 B.C.

THE ORBITS GET BETTER

Not long after Halley's achievement firmly established that comets can return, Anders Johan Lexell showed in the 1770s that Comet Messier 1770 I (the Roman numeral I means that it was the first comet to round the sun in 1770) was also a periodic comet. It has a period of only 5 years. However Lexell also showed that this wayward comet had come close to Jupiter just before its approach to the Earth and that afterward it came close to Jupiter again. In this second encounter, Jupiter changed the orbit so drastically that the comet would never return to the vicinity of Earth.

As our story gets closer to the present, we come across other ingenious attempts to link one comet appearance with another; some were successful, but others, announced with far more hoopla, were not.

Early in the nineteenth century, Johann Franz Encke became curious about several comet appearances. In 1786 France's Pierre-François Méchain discovered a comet, and Caroline Herschel of England found one in 1795. From his observing sites in France, Jean-Louis Pons found two more in 1805 and 1819. Using accurate observations to calculate their orbits, Encke found that they fit the orbit of a single comet that would next return in 1822. Encke was right, and the comet that now bears his name has by far the shortest orbital period of any comet, a brief 3.3 years.

During the last week of October 1366, a large comet crossed the sky from Ursa Major to Aquarius and was well recorded by Chinese observers. In the mid–nineteenth century, J. Russell Hind suggested that this large comet might be an early appearance of Periodic Comet Tempel–Tuttle, the parent of the Leonid meteor stream. In the early 1960s, John Schubart did the calculations that linked the two comets and found the record of Gottfried Kirsch's single observation (never confirmed by another observer) of Temple–Tuttle in 1699. Schubart predicted that the comet would return in 1965; his prediction was off by only 5 days.

THE COMET OF CHARLES V

Hind was not so lucky with his ideas about a possible return of the comet of 1556, the comet that marked the end of the long reign of Charles V of the Holy Roman Empire. Charles was the Hapsburg emperor who ruled more territory than anyone else before or since. Let others wage war, was his philosophy about expanding his rule. "You, lucky Austria, marry." He retired of his own will around 1556, living quietly on his estate and indulging his interest in clocks, and died 2 years later.

Charles's comet appeared in the southern sky of 1556, moving through Corvus and Virgo. In 1751 Richard Dunthorne suggested that it was the same comet "that created astonishment throughout Europe" in 1264.[10] He went on to prophesy a return in 1848; in 1857 with Charles's comet still unseen, Hind refined that forecast to August of either 1858 or 1860.

In the following 3 years, two very bright comets appeared, Donati's in 1858 and the great comet of 1861. "The new visitor," trumpeted the London *Times*, "which has taken even astronomers by surprise, shone with great brilliancy last night [July 3, 1861], exciting universal admiration. . . . Its size does not at present exceed that of 1858, and it differs from it materially in this respect, that its tail is straight instead of being curved."[11] A competing newspaper described a first view of the comet on June 30: "It then far exceeded in brightness any comet I have observed, those of 1811 and the recent splendid one of 1858, not excepted."[12] All this excitement naturally launched a debate as to whether the 1861 comet might be that of Charles on its long-awaited return. But the original assumption was all wrong: Charles's comet was not the comet of 1264, and neither comet has ever reappeared.

Even in Hind's day, astronomers understood that gravitational pulls of the planets can affect the period of a comet as it travels about the sun. "It is true," he wrote in 1859,

> if a comet experienced no resistance while performing its journey round that luminary, it would make its appearance after equal intervals of time. . . . But the movements of comets are greatly disturbed by the attraction of the various planets belonging to the solar system, particularly by Jupiter and Saturn, which far exceed the rest in magnitude."[13]

Hind even addressed the possible effects of a collision with Charles V's comet: "At worst," he grossly underestimated, "a direct collision would perhaps be comparable only as regards the mechanical effect upon the earth to a meeting with a huge cushion."[14]

Early this century the English astronomer Andrew C. D. Crommelin fit the orbits of comets found by Pons, Coggia, Winnecke, and Forbes into that of a single returning object. Just as the earlier Comets Halley, Lexell, and Encke were named to honor the accomplishments of those who had unravelled their mysteries, this newest comet was renamed Periodic Comet Crommelin.

Some 250 years after Halley showed us how to do it, figuring out where and when a comet will return involves the same difficult mathematics. The result is every bit as exciting as discovering a comet, even though computers compress months of calculations into seconds. "In 1973," writes Marsden, "I had the benefit of a computer for testing out the effect of assumption of slightly different initial revolution periods."[15]

One can argue that Marsden's achievement with Comet Swift–Tuttle is in the same class as those of Encke and Crommelin, even though his computer made his work go faster. To keep up with the tradition, would it not be a good idea to change the name of Comet Swift–Tuttle to Marsden? Maybe not, some people might say; after all is it practical to rename a comet we have called Swift–Tuttle for 130 years? But Periodic Comet Marsden has a nice ring to it. Whether or not the name is changed, I hope that our descendants will look at the comet of 2126 as it surges by and say, "There goes Marsden's comet." It would be a well deserved tribute to him.

Comet or Planet?

Friedrich Wilhelm Herschel would become one of the greatest figures in the history of astronomy, but in 1760, he was an impoverished musician who didn't have the funds to return to England from a trip to Genoa, Italy. Desperate to get home, he gave a bizarre impromptu recital—a maestro playing a harp, holding a horn, and having a second horn attached to his shoulder. It may be hard to picture Herschel playing all three instruments in such a pose, but the people of Genoa came in droves, and he made his way back to England.

Born in Hanover, Germany, in 1738, Herschel was the son of a musician, and he became a composer and conductor as well as a skilled performer on horn, harp, and organ. He used the English version of his first name after he moved to England in 1757. At this period of his life, Herschel's highest ambition was to establish his reputation in music. To get from place to place to play, he bought a horse that he frequently rode across the fields in typically damp and stormy British weather. In addition to his extraordinary schedule, he packed his evenings with education: During all this time, he wrote,

> Though it afforded not much leisure for study, I had not forgot
> my former plan, but had given all my leisure hours to the

study of languages. After I had improved myself sufficiently in English, I soon acquired the Italian, which I looked upon as necessary for my business. I proceeded next to Latin, and having also made considerable progress in that language, I made an attempt of the Greek . . . but soon dropped the pursuit of that as leading me too far from my other studies, by taking up too much of my leisure.[1]

These other studies, Herschel goes on, were in music:

The theory of music being connected with mathematics, had induced me very early to read in Germany all that had been written upon the subject of harmony; and when, not long after my arrival in England, the valuable book of Dr. Smith's harmonics came into my hands, I perceived my ignorance and had recourse for other authors for information, by which I was drawn from one branch of mathematics to another.[2]

MUSIC OF THE SPHERES

In May 1773 Herschel made a small purchase that would have major consequences. He records in his diary, "May 10. Bought a book of astronomy and one of astronomical tables."[3] Within a few weeks, his passion for the order and harmony of music had transformed itself into a passion for the sky, and he plunged into his new love with his typical zeal. Buying some lenses, he built a 4-foot-long tube and mounted a lens in it: "With this," he exulted,

I began to look at the planets and the stars. It magnified 40 times. In the next place I attempted a 12 feet one and contrived a stand for it. After this I made a 15 feet and also a 30 feet refractor and observed with them.[4]

Although Herschel enjoyed these refractors, he found them long and unwieldy. Based on Galileo's 1609 design, they involved a single objective lens that bent the light so that it would reach a focus at the eyepiece end of the telescope, down at the other end of the tube. Moving a skinny 30-foot-long tube across the sky was a challenge, especially if the tube failed to stay absolutely straight to allow the light to travel down its length to the eyepiece. Perhaps, he thought, the reflector telescope that Isaac Newton had invented in 1672 by putting a concave mirror at the bottom of a tube and an eyepiece near the top would better suit his needs. So around September 8, Herschel completed a 2-foot-long Gregorian-type reflector, in which a concave secondary mirror sends light through a hole in the main mirror to the eyepiece.

> This was so much more convenient than my long glasses that I soon resolved to try whether I could not make myself such another . . . I was, however, informed that there lived in Bath a person who amused himself with repolishing and making reflecting mirrors.

When Herschel arrived however, he found the optician's interest waning. The craftsman immediately offered Herschel "all his tools and some half-finished mirrors, as he did not intend to do any more work of that kind. . . . About the 21st October I had some mirrors cast for a two-feet reflector." Following the custom of the day, Herschel did not use glass but a form of speculum metal that consisted, he wrote, "of 21 copper, 13 tin, and one of Regulus of Antimony, and I found it very good, sound white metal."[5] Unlike today's glass objectives, speculum mirrors do not need additional reflective coatings, and telescopes over 200 years old still survive, their fragile coats of metal still longing for a skyward look. By the beginning of November, Herschel had become so proficient in completing mirror after mirror that he was getting behind in finishing the telescopes that the mirrors would serve.

So thoroughly did Herschel immerse himself in his zeal for telescopes that he needed a larger house in Bath. His younger sister Caroline Lucretia, born in 1750, moved into it as well, joining him enthusiastically in his work and not minding all the telescopes strewn about in various stages of construction. The backyard of their house was set up as an observing site. About the size of a modern backyard, it had room for several small telescopes.

A MAJOR DISCOVERY

By 1781, only 8 years after he fell in love with the stars, Herschel was deeply involved in a systematic survey of the sky. "On Tuesday the 13th March," he wrote,

> Between ten and eleven in the evening, while I was examining the small stars in the neighborhood of H Geminorum, I perceived one that appeared visibly larger than the rest; being struck with its uncommon magnitude I compared it to H Geminorum and the small star in the quartile between Auriga and Gemini, and finding it so much larger than either of them, suspected it to be a comet.[6]

Although the object was pretty bright, perceiving it not as a point of light but a disk was a real challenge even to a trained eye. Herschel later wrote,

> Seeing is in some respects an art which must be learnt. To make a person see with such a power is nearly the same as if I had been asked to make him play one of Handel's fugues upon the organ. Many a night have I been practising to see, and it would be strange if one did not acquire a certain dexterity by such constant practice.[7]

News of this peculiar comet spread swiftly. Across the English channel, Charles Messier, a French astronomer who would become equally revered, observed the new object at every opportunity. Messier was stumped by both its ultraslow movement from night to night and its shape. "I am constantly astonished at this comet," he wrote to Herschel in the late spring of 1781,

> which has none of the distinctive characters of comets, as it does not resemble any one of those I have observed, whose number is eighteen . . .

> I have since learnt by a letter from London that it is to you, Sir, that we owe this discovery. It does you the more honor, as nothing could be more difficult than to recognize it, and I cannot conceive how you were able to return several times

to this star—or comet—as it was absolutely necessary to observe it several days in succession to perceive that it had
motion . . . .

For the rest this discovery does you much honour; allow me
to compliment you for it. I should be very curious, Sir, to learn
the details of this discovery, and you will oblige me if you
will be so good as to inform me of them.[8]

"Life has been one long press conference," Marsden lamented
in 1992 about the fuss with his successful prediction of the return
of Comet Swift–Tuttle and its possible near encounter with the
Earth in 2126. More than two centuries earlier, Herschel was just
as annoyed at the mounting publicity about his comet. He was
trying to build a mirror for a new 20-foot-long reflector. If a minute
could be spared from all the publicity, he would undoubtedly get
back to it.

On August 31, 1781, Anders Lexell, the mathematician who
had determined the orbit of a comet, blew the whistle on Herschel's
claim that his remarkable object was a comet. Its orbit was almost
circular, he announced, and it never gets closer to the sun than 16
times the Earth's distance to the sun. The pieces of the puzzle had
finally fallen into place. Herschel's object did not look much like
a comet because it was not a comet. It was a planet, the first to be
discovered in historic times.

With the confidence of a man who once played solo with a
harp and two horns, Herschel suggested that his methods were so
precise that he could not possibly have missed finding the new
object, whether it turned out to be a comet or a planet. Standing
toward the top of his instrument, he would sling it to and fro,
observing stars as they went through the field of view and noting
their positions, colors, and brightnesses. Once the new find was
identified as a planet, the politically astute Herschel wanted to
name it Georgium Sidus after his patron, King George III, but it is
now called Uranus.

At the end of 1781, the Royal Society offered Herschel its
Copley medal. The uncannily prescient citation is worth noting:

> Your attention to the improvement of telescopes has already
> amply repaid the labour which you have bestowed upon them;
> but the treasures of the heavens are well known to be inex-
> haustible. Who can say but your new star, which exceeds
> Saturn in its distance from the sun, may exceed him as much
> in magnificence of attendance? Who can say what new rings,
> new satellites, or what other nameless and numberless phe-
> nomena remain behind, waiting to reward future industry?[9]

Almost two centuries later in 1977, Uranus passed in front of a
star. The star blinked out several times before Uranus occulted it,
then the star repeated its performance the same number of times
afterward. Thus a set of rings was found about Uranus. More rings
were found when *Voyager II* sailed past Uranus as part of its od-
yssey through the outer solar system. The planet is known to have
15 satellites, two of which Herschel discovered himself.

After the German-born astronomer was appointed private astronomer to the king in 1782, George III insisted that Herschel
have the best telescope possible. An avid amateur astronomer
himself, George III had established the Kew Observatory near the
present site of the Kew Botanical Gardens. The first meeting at
Windsor castle between George III and Herschel was a great success, and Herschel was delighted with their observing session on
the evening of July 2: "My Instrument gave a general satisfaction,"
he wrote to his sister Caroline, "the King has very good eyes and
enjoys Observations with the Telescopes exceedingly."[10] With the
king's munificence Herschel continued to indulge his love of telescope making. In 1787 he redesigned his 20-foot telescope, taking
out the diagonal and sending the light at an angle directly to the
eyepiece, a design now called the Herschelian telescope. By 1800
he had made some 80 20-foot-long reflectors and double that
number of 10-foot-long telescopes.* By the end of his life, Herschel
had made an estimated 2160 telescopes, few of which survive.

---

* In Herschel's day the focal length of the telescope, the distance between the
mirror and the focused eyepiece, was the prime way of referring to a telescope.
Today we first discuss the diameter of the mirror or the lens.

THE COMETS OF CAROLINE

Herschel's self-discipline and enthusiasm were infectious, especially for his sister Caroline, who became a famous astronomer in her own right. In the early 1780s, she began hunting for comets using a 6-inch diameter reflector her brother had built for her work. By 1783 she had discovered the beautiful spiral galaxy in Sculptor now known as NGC 253. On August 1, 1786, when Herschel was in Germany, she reported finding a new comet, an act that thrilled her brother but got her a somewhat condescending official reply:

> Let us hope the best and that it is approaching the earth to please and instruct us and not to destroy us, for true Astronomers have no fears of that kind, witness Sir Harry Englefield's valuable tables of the apparent places of the comet of 1661 . . . I would not affirm that there may not be some astronomers so enthusiastic that they would not dislike to be whisked away from this low terrestrial spot into the higher regions of the heavens by the tail of a comet.[11]

After Caroline's comet finds became more routine, she was awarded a small income, and observatory replies were more professional and dignified. The Royal Astronomical Society bestowed its gold medal on her in 1828.

Caroline discovered eight comets in all. Her second find, at the end of 1788, turned out to be periodic, returning about every 150 years. When Roger Rigollet discovered a comet in 1939, the celestial mechanician Leland Cunningham determined that it was Herschel's comet returning; it is now known as Periodic Comet Herschel–Rigollet. Another comet turned out to be an appearance of Periodic Comet Encke. Caroline found two comets in 1790 and another in December 1791. By the time she had found her eighth comet in 1797, England's most famous woman astronomer was known throughout Europe for her "eccentric vocation" that included a log of her discoveries lovingly entitled "Bills and Receipts of my Comets." Despite her success with comet hunting, Caroline insisted that her greatest contribution was in assisting her brother.

With two large grants totalling 4000 pounds from King George, Herschel built a mammoth 40-foot-long telescope with a 48-inch diameter mirror. From the first night of its use, when he discovered Mimas and Enceladus, two of Saturn's closest moons, Herschel knew that this was a superb instrument. However its large mirror tarnished quickly, and since the telescope was unwieldy, it spent more time under repair than under the stars. Extremely difficult to use, the telescope's eyepiece could be reached only by standing on a high platform and shouting orders to an assistant.

Caroline had a bad accident on the telescope one night when her ankle was struck. As the story goes, one night Herschel asked her to move the telescope quickly in a particular direction. When nothing happened, Herschel repeated, "Lina, move the telescope!" But a large hook used for pulling the telescope had snagged her ankle and Caroline yelled back "I'm hooked!" As soon as she was almost recovered from this injury, she returned to assist her brother in his work.

In May 1788 Herschel married Mary Pitt. "His wife seems good-natured," Fanny Burney, a visiting novelist, wrote about her. Then noting that she was a widow with a large fortune, she added cattily that "astronomers are as able as other men to discern that gold can glitter as well as stars."[12] The Herschels had a son, John Frederick William, who first distinguished himself as a mathematician but later followed his father's footsteps to the stars. He went to the Cape of Good Hope in 1834 where he discovered and measured many previously unseen clouds of gas and clusters of stars in the southern sky. After he returned from his survey of the southern sky in Africa, John found the 40-foot telescope in poor shape. The scaffolding was weak and dangerous, and with considerable sadness, he decided to dismantle it. Years later lightning hit a tree, which fell on what was left of the huge telescope. All that now remains of Herschel's great 40-foot telescope is the bottom stub from the tube. In a requiem for New Year's 1840, the Herschel family met at the telescope to recite a poem John had written:

In the old Telescope's tube we sit,
And the shades of the past around us flit;
His requiem sing we with shout and din
While the old year goes out, and the new comes in.

> *Merrily, merrily, let us all sing,*
> *And make the old Telescope rattle and ring!*

God grant that its end this group may find
In love and in harmony fondly joined!
And that some of us, fifty years hence, once more,
May make the old Telescope's echoes roar.

> *Merrily, merrily, let us all sing,*
> *And make the old Telescope rattle and ring!*[13]

William Herschel's accomplishments were prodigious. He determined the rotation period of Saturn to be a rapid 10 hours. From his careful records of the positions of more than 800 double stars over many years, he concluded in 1803 that some pairs revolve around a common center over a period of years. He multiplied Messier's catalog of nonstellar objects many times, observing and listing more than 2500 deep-sky objects called nebulae. But he loved his telescopes even more than what he could see with them. "It would be hard to be condemned," he wrote,

> because I have tried to improve telescopes and practised continually to see with them. These instruments have played me so many tricks that I have at last found them out in many of their humours and have made them confess to me what they would have concealed, if I had not with such perseverance and patience courted them. I have tortured them with powers, flattered them with attendance to find out the critical moments when they would act, tried them with specula of short or long focus, a large aperture or a narrow one; it would be hard if they had not been kind to me at last.[14]

❇ 5 ❇

The Sport Begins

*C*omet hunting is by far the slowest of all sports. It is one long game, actually, that began with Messier and is still in full swing. The playing field is worldwide, and players share the competitiveness, the egos, and the envy that electrify other sports. Cutting across class lines, each contender has reasons for finding a new comet, whether it be fame, winning, or just an interest in comets. Japan's Kaoru Ikeya was a worker in a piano factory; Canada's Rolf Meier is an electrical engineer; and Ohio's Leslie Peltier designed juvenile furniture. Howard Brewington, who lives with his wife, Trudy, near a small mountain town in New Mexico just for the dark sky that comet hunting requires, repairs television sets for a living.

Messier was certainly not the first person to find comets; every comet seen was discovered by somebody, although early records are lost. Paolo Toscanelli, the late fifteenth-century cartographer whose map of the Atlantic Ocean encouraged Columbus to set out on his voyages, also made careful drawings of six bright comets that appeared between 1433 and 1472. A man of eclectic interests besides cartography and astronomy, Toscanelli may have observed the sky so often that he saw these comets before others told him about them, but maybe he did not. Centuries later it does not matter. What is important is that his drawings allowed later astronomers to compute their orbits, including that of Halley's comet, whose visit in 1456 frightened Pope Calixtus.

THE COMET FERRET

I suspect that Messier got his comet-hunting start thanks to his failure to be the first person to see Halley's comet in 1758. Perhaps he wouldn't admit it, but I like to imagine that he decided to get even with the sky, so to speak, by finding every other new comet that appeared after he was beaten out by Johann Georg Palitzch, a farmer from Dresden. But Messier's relationship with Halley's comet is more complicated still, for when he found Halley he had not even heard of Palitzch, news of whose find somehow did not reach Paris for several months. On January 21, 1759, while working at Nicholas Delisle's observatory at the Hôtel de Cluny in Paris, Messier finally saw the comet. He was thrilled: "It was one of the most important astronomical discoveries," he wrote, "for it showed that comets could return."[1] He summoned Delisle, who observed the comet, and then promptly ordered Messier not to announce it in any way. Messier complied with the request, so the comet was not announced until April 1, 3 weeks after it already rounded the sun. By this time, Messier had already heard of Palitzch's Christmas night sighting and knew that he would have lost the race anyway despite Delisle.

Messier didn't have long to wait to assuage his disappointment. Discovering his first comet in 1760, he made his first eight finds with the playing field virtually to himself. He found comets in 1763, 1764, 1766, 1769, 1770, 1771, and 1773 before his chief rival, Jacques Montaigne, a druggist in Limoges, France, snared his first in 1772, the comet that in 1826 became known as Biela's Periodic Comet or P/Biela. This comet apparently split in two in 1846, returned as two comets in 1852, and only as a meteor storm in 1872. By 1781 Pierre Méchain had also joined the contest. But in 1801 Messier was still in the lead, having added six more notches to his telescope.

The most important commodity needed by Messier and every comet hunter that followed him is patience. Hunting for comets, as we've noted, demands time—not time set by the hunter but by the sky itself. The moon acts as a referee, brightening the sky and limiting the period available for searching. The game is especially competitive just after full phase. For the preceding week, the waxing moon swathes the evening sky in light that makes all but the brightest comets invisible. A comet could brighten slowly during that week under the protective cloak of the moon's light, unnoticed by searchers. But the moon continues its eastward trek across the sky, rising later each night. Within a day or two after the full moon, the evening sky is suddenly thrust into darkness, and comet hunters all over the world rummage through it in a scramble for new comets. Some 10 days later, an almost-new moon opens the morning sky for another inning as sleepy searchers peer through their telescopes to see if anything new has appeared.

Except for the added dimension of photographic finds, the game has changed little in 200 years. A dark country sky is the same sky it was in Messier's time, with one big exception—the French sky gazer would have been amazed to see that the Earth has not one moon but hundreds of them, all artificial. These are the satellites that have appeared in increasing numbers since 1957, all looking like wandering stars moving gracefully across the telescope's field of view. Since Messier observed from the middle of Paris—he noted the passage of every hour by bells from more than 40 churches—he would have known something about light pollution, the bane of big-city sky gazing. Industrial pollutants and lights brighten up the sky, drowning out all the fainter stars and all but the brightest galaxies, clusters, and comets.

Messier's methods and telescopes are similar to what searchers employ now. He used a low power 2-inch diameter refracting telescope, though later he switched to a larger instrument. He moved his telescope steadily back and forth through a region of sky, looking for a fuzzy patch of light that crept so slowly he needed an hour to notice if it had moved at all.

In 1986 while visiting the Paris Observatory, I perused one of Messier's observing logs. As I studied his notes, his descriptions of the telescope he used, his complaints about the weather, and his comments about what he was looking at, 200 years seemed to slip away. Not that much has changed; reflecting the same concerns and joys, his notes could well have been my notes. It is still the same game.

So many comets found their way into Messier's telescope that Louis XV called him a ferret of comets, and President Jean Baptiste de Saron of the Paris parliament, a man versed in comet orbit calculation as well as politics, asked for observations of all Messier's comets so that he could calculate their orbits. Messier happily complied. The two men became great friends. One evening de Saron and his family brought Messier to an oriental garden at the Parc Monceau that had many small pagodas, castles, and passageways. Somehow Messier became separated from de Saron. Seeing a small open door, Messier, curious as always, stepped through. There was nothing inside but a long drop to a storeroom of ice. Messier fell 25 feet, splitting his scalp and breaking an arm, a leg, and several ribs as he crashed onto the ice below. Messier's injury kept him out of commission for almost a year, but by 1782, he was back at the telescope, and in 1785 he discovered another comet.

But the next decade was a difficult one for Messier. The French Revolution saddened and frightened him during the last years of his search. With his social peers literally losing their heads, the gentleman astronomer feared for his own safety. Forced to leave the observatory in Paris, he continued his work in a quieter location. On the evening of September 27, 1793, Messier found a comet in Ophiuchus, and like so many times before, he informed his friend de Saron, who attempted to calculate an orbit using the positions Messier supplied. Although this comet was independently discovered by Caroline Herschel, a comet hunter from England, it was visible for only a few nights before it sank into the evening twilight.

However de Saron was no longer president of the Paris parliament. Accused of being an enemy of reform, he was in prison and awaiting execution by that creature of the Revolution, Madame

Guillotine. It is hard to imagine that de Saron could have cared about comet orbits when he was about to lose his head, but he did manage to calculate an orbit for Messier's comet from his prison cell. If de Saron's orbit was correct, the comet would come closer to the sun, then move away, and reappear in the morning sky.

On December 29 Messier searched the morning sky and found his comet close to the position de Saron had predicted. Messier wrote of de Saron's last success and hid his note in a newspaper that he was able to smuggle to the prisoner. On April 20, just 3 months before the end of Robespierre's Reign of Terror, de Saron was guillotined.

Although Messier survived the revolution and the Reign of Terror that followed it, his pension, which de Saron had arranged for him after his accident, was gone, and the acclaimed astronomer was virtually penniless. Messier did find another comet in 1798. There is a well-known but unverified story from his Russian friend, La Harpe, that Messier found out about a discovery by Montaigne while mourning his wife's death. A friend embraced the grief-stricken man to say, "I am so sorry." Messier glared at his visitor. "Alas," he said, "Montaigne has robbed me of my comet!" Quickly realizing his error, Messier tried to recover. "Poor woman," he muttered. No doubt his friend agreed.[2]

COMET MASQUERADERS

As the first serious comet hunter, Messier learned about the sand traps of the sport in time to warn the rest of us. Fuzzy objects that are not comets lurk all over the sky, waiting to snare naive observers. Peltier called them comet masqueraders. In fact, according to Daniel Green of the CBAT, approximately 98 percent of comet discovery reports from new observers turn out to be spurious. Most of these ecstatically reported sightings turn out to be galaxies, star clusters, or clouds of gas called nebulae. At the end of 1758, when Messier found a fuzzy patch near the star Beta Tauri that never moved, he began to catalog these deceivers.

The first entry in his catalog, now called Messier 1 or M1, is more popularly known as the Crab Nebula because it resembles a ghostly version of the sea animal. Although Messier never realized it, Messier 1 was probably the most interesting thing ever to cast light into his telescope. It is the remnant of the supernova, a near-total destruction of a star, observed on July 4, 1054, to be as bright as Venus. A pre-Columbian petroglyph in northern New Mexico depicting a star near a crescent moon may be a record of this stellar catastrophe. In only 700 years, the explosion has produced a huge expanding shell of gas bright enough that Messier could easily see it. In 1969 observers using a 36-inch diameter telescope at Steward Observatory in Arizona found an object at its center, a pulsar that spins 30 times per second.

Besides the supernova remnant, Messier's catalog includes the best and brightest of the northern hemisphere sky's interesting fuzzies. Some of these Messier discovered; others he merely listed. Observing all of them has become the goal of many amateur astronomers. In 1962 I began my own Messier hunt with a single observation of the Pleiades, M45. (Why Messier included this big, bright cluster, which does not look at all like a comet, is a bit of a mystery.) In spring 1967, using a larger telescope than I had in 1962, I finished my list while observing from my grandfather's cottage at Jarnac Pond, Quebec.

Messier published three versions of his catalog. The first 45 objects appeared in 1774, and by 1781, his list had grown to 103. Besides the roster that Messier created, other objects he recorded but never listed were later added, so that now the Messier catalog includes 110 objects spread over much of the sky.

LE GRAND CHERCHEUR

Messier discovered his last comet on July 12, 1801. However a 40-year-old janitor beat him to it, finding the comet near the Big Dipper a day earlier and receiving full credit for the find. This little comet was the first of 26 comets that now bear the name of Jean-Louis Pons.

Actually Pons had more than 26 comet discoveries, and perhaps as many as 37.[3] Born in 1761, Pons didn't become interested in the sky until 1789, when the observatory at Marseilles, France, hired him as a doorkeeper to keep watch on the grounds. He did, and kept watch on the heavens as well.

Pons was justifiably proud of le grand chercheur—his favorite searching telescope. With a 3-degree field of view (six full-moon diameters), he could cover broad swaths of sky at once. Pons found a second comet in 1802, one in 1804, and another in 1806. While hunting with this telescope just before dawn on February 9, 1808, Pons saw a comet very close to the globular cluster Messier 12. However, Pons typically kept rather poor records of his observations. In a field crowded with stars, the only three objects he drew were the comet, Messier 12, and the nearby globular cluster Messier 10. Since no one else was able to observe this object, no orbit was ever calculated for it, and for some 180 years, the comet remained unconfirmed. Not long ago Pons's 1808 comet was revealed to be an early passage of Periodic Comet Grigg–Skjellerup, a faint comet that returns every 5 years, so this ancient story is now complete.

Despite his growing record of comet finds, Pons was rather slow to gain the respect of his peers, possibly because his rural upbringing made him appear unsophisticated. According to a story told years later by the astronomer Augustus De Morgan, after Pons had gone some time without finding anything, he asked the director of the Seeberg Observatory, Baron von Zach, for a hint on how his comet sweeping could be more productive. Search when there are lots of sunspots on the sun, the German astronomer suggested as a practical joke. Half-expecting that Pons would make a fool of himself and redesign his program to search the night sky when the sun sported large spots, von Zach was very surprised to get a letter from Pons with profuse thanks. Large spots indeed formed on the sun, and soon afterward he dutifully found a new comet.[4] To this day no correlation between comets and sunspots has been found.

In 1811 Pons discovered a comet destined to be one of the best and brightest in history, however he found it 3 weeks after the French observer Honoré Flaugergues. The comet quickly brightened until it could be seen without a telescope, and it remained visible to the naked eye for 10 months. Its appearance was even credited with the coincidentally ultrafine wines that year. In October 1816 the English poet John Keats wrote of the thrill of reading a new work of literature:

> *Then felt I like some watcher of the skies*
> *When a new planet swims into his ken;*
> *Or like stout Cortez when with eagle eyes*
> *He star'd at the Pacific—*[5]

Most critics assume that Keats was recalling Herschel's discovery of the planet Uranus in 1781, an event that occurred 14 years before he was born. I suspect the line harks back to the discovery by Pons and others of the magnificent comet of 1811, which appeared only 3 years before he wrote the sonnet.

In 1813 with growing success as a comet hunter, Pons was promoted from doorkeeper to assistant astronomer, and in 1817, he became director of an Italian observatory near the town of Lucca, not far from Florence, a site from which he discovered seven more comets. In the twilight of his career, Pons became director of Florence's Observatory and Museum for Physics, where he spied another seven comets. He found his last comet in August 1827, and he died 4 years later. Only one person, Carolyn Shoemaker, has beaten Pons's record of 26 named comets.

In 1831 Denmark's Frederick VI added some spice to the comet-hunting game by striking a gold medal for each discoverer of a comet. It was the first of several comet-finding medals given in different countries and times, and it is this medal that gave rise to the tradition of naming comets after their first discoverers. To take up the slack left when this award ended in 1848, the Vienna Academy of Sciences honored each comet finder with a prize; this award lasted another 30 years. The most lucrative of all the comet prizes was given by H. H. Warner to any U.S. discoverer of a comet and helped spur a golden age of hunting during the late nineteenth century.

THE COMET PENDULUM SWINGS WEST

By the middle years of the nineteenth century, the bulk of comet discoveries had moved across the Atlantic Ocean, where Lewis Swift and Horace Tuttle were competing for new comets.

By the 1880s a new generation, led by William R. Brooks and Edward Emerson Barnard, had taken over. With formidable career totals of 22 and 16 named comets, respectively, these hunters found their first comets within a month of each other and were keen competitors.

Born in Scotland, William Brooks moved with his family to New York in 1857, settling upstate in Phelps, where he became a photographer. Brooks made the 9.25-inch reflecting telescope with which he searched. Watching the dazzlingly bright Donati's comet as it made its way past Earth in 1858 had a deep impact on the young man, although it wasn't felt right away.

In the 1870s he started to build his own telescopes. With a 5-inch reflector, he independently discovered Comet 1881 V in October 1881, not long after the British comet hunter William Denning first spotted it.[6] A new periodic comet, 1881 V, faded rapidly after discovery, and it was not seen at its predicted return in 1890. However in October 1978, the Japanese comet hunter Shigehisa Fujikawa rediscovered the comet, and it was renamed Denning–Fujikawa. Again the comet was not seen on its next return in 1987.

Edward Emerson Barnard was born in Nashville, Tennessee, in 1857. He got an early start in optics: When he was less than 10 years old, he took a job as a photographer's assistant after leaving school with but 2 months formal education. Reading Alexander Ewing's *Practical Astronomy* in 1876 ignited his interest in the heavens.[7] With his first telescope, his main love became Jupiter. His intimate knowledge with the giant planet paid off when years later as an astronomer at Lick Observatory, he discovered Jupiter's fifth moon Amalthea, the first Jovian moon spotted since Galileo's discoveries in 1610. Less than a month before Brooks found his first comet, Barnard found his on September 17, 1881.

There is no question that the Warner prize propelled the competition for new comets during this period, for Barnard so much as admitted it: "I had been searching for comets for upward of a year with no success," he writes,

> when a prize of two hundred dollars for the discovery of each new comet was offered by the founder of the Warner Observatory through the agency of Dr. Lewis Swift, its Director. Soon after this it happened that I found a new comet, and was awarded the prize. Then came the question, "What shall I do with the money?" After due deliberation it was decided that we would try to get a home of our own therewith. I had always longed for such a home, where one could plant trees and watch them grow up and call them our own. So we bought a lot with part of the money, which was on rising ground selected in part because it gave me a clear horizon with my telescope.
>
> After some saving and mainly a mortgage on the lot, we built a little frame cottage where my mother, my wife and I went to live. Those were happy days, though the struggle for a livelihood was a hard one, working from early to late, and sitting up the rest of the twenty-four hours hunting for comets. We looked forward with dread to the meeting of the bills which must come due. However, when this happened, a faint comet was discovered, and the money went to meet the payments. The faithful comet, like the goose that laid the golden egg, conveniently timed its appearance to coincide with the advent of those dreaded notes. And thus it finally came about the house was built entirely of comets. This fact goes to prove the great error of those scientific men who figure out that a comet is but a flimsy affair after all, infinitely more rare than the breath of the morning air, for here was a strong compact house, albeit a small one, built entirely of them. True, it took several good-sized comets to do it, but it was done, nevertheless.[8]

Combining the best of a new marriage and new comets, Barnard's playful words capture the enthusiasm with which he followed his mission. When I first read them, I began to look for some small connection to my own program. At the time I had zero comets, so I really had to pull at straws. Moreover his house was under a dark country sky; mine was in a big city. Finally I learned that his house was on Belmont Street in Nashville, and I lived on Upper Belmont Avenue in Montreal. There!

A DECADE FOR COMETS

Barnard had something else going for him. The decade of the 1880s brought one of the most glorious series of bright comets to parade through the inner part of the solar system. The onslaught of comet after major comet must have spurred on Barnard and his competitors to greater glories. Compared especially to the relatively paltry crop we have had in recent years, the procession was truly splendid. It began in 1880 when a large comet dominated the southern sky constellations of Tucana, Grus, and Phoenix; its tail grew to 40 degrees in length to cover a quarter of the visible sky. On May 22 of the following year, the Australian comet hunter John Tebbutt discovered a comet that brightened to naked-eye visibility and dominated the night sky for 2 months.

By the following May, the first great comet of 1882 was a bright naked-eye object, brighter than Jupiter and visible in a twilight sky. Four short months later, the great September comet brightened to magnitude minus 14—much brighter than the full moon—as it reached perihelion in a hairpin dance around the sun. David Gill, director of South Africa's Cape Observatory, described the sight of the comet's rising:

> There was not a cloud in the sky, only a merging into a rich yellow that fringed the blackish blue of the distant mountains, and over the mountains and amongst the yellow an ill-defined mass of golden glory rose with a beauty I cannot describe.

> The sun rose a few minutes afterwards, but to my intense
> surprise the comet seemed in no way dimmed in brightness,
> but becoming instead whiter and sharper in form as it rose
> above the mists of the horizon.[9]

The excitement this comet generated was so intense that when
Barnard went to sleep early in the morning of October 14, he dreamt
about comets—lots of them, comets all over the place. The story
was so real to him that when he awoke some hours later, it took
a few minutes before he realized it was all a dream.

Or was it? The sky was still dark, still clear, and with the
mighty comet just rising in the east, Barnard rushed outside to his
waiting telescope. Turning it to the comet, he studied its marvelous
structure before beginning a scan for new comets. Moving his telescope to the southwest, he didn't have far to go—perhaps 5 or 6
degrees—before he found a group of a half-dozen small comets.
Flabbergasted, Barnard wondered whether he had fallen asleep
on his feet and had resumed his dream. The comets were real—
the observation was confirmed the following night by observers
in Europe. But although the little comets were traveling at the same
rate and direction as the main body, they all disappeared in less
than a day.

Brooks did well with the great comet, too. A week after Barnard's find, on October 21 he found a companion comet several
degrees northeast of the comet. Brooks's object also vanished. Although the main comet broke into several pieces at perihelion—a
comet's most vulnerable time—on September 17, these pieces
stayed within the coma of the main object. The origin of these
objects viewed by Barnard and Brooks is a mystery; they could not
have split off the main comet at perihelion and moved so far from
it in such a short time.

A parade of sun comets that hit the sun a century later may
offer a clue to this mystery. In the 1980s the Solwind and Solar
Maximum Mission satellites discovered what appeared to be comets
that hit the sun. Some of these came in groups, notes Marsden,
one pair only 12 days apart. Speculating that it is possible that the
Barnard and Brooks comets split off from the main comet when it

last rounded the sun, Marsden suggests that they might have rounded the sun unseen in September 1882 and then flared for a few hours, long enough to be seen by these comet sleuths.

Since the next big comet would not appear for 3 years, Brooks, Barnard, and their fellow searchers had a little break. The comet that came along at the end of 1885 reached second magnitude—about the brightness of the North Star—and was followed by a fainter comet that Barnard found a year later. Brooks had a banner year in 1886: In a remarkable 4-week period, he discovered three comets, one on April 27, another May 1, and a third on May 22.

Brooks's triple success was followed by a Barnard threesome—he discovered comets on October 4, 1886, and on February 16 and May 12, 1887. But it was a farmer near Cape Town who walked away with 1887 I, by far the brightest comet in those 2 years. Already visible to the naked eye, as it grew brighter it developed, as one observer noted, a brilliant silver white tail stretching a quarter of the way around the sky. Moreover in 1888 another bright comet came by. Finally as a close to this prolific decade, Barnard found a comet early in 1888 that did not reach perihelion until a year later. Cruising slowly around the sun, this comet was visible for almost a thousand days.

THE HOAX OF THE COMET-SEEKER TELESCOPE

With increasing success and fame, Barnard made the transition from amateur to professional by taking a position at Lick Observatory near San Francisco in 1888. By now he had found so many comets that a competitor, disgusted after losing comet after comet to Barnard, asked him why he couldn't keep them on a leash. Barnard might have felt the same way about reporters when on March 8, 1891, he saw this astonishing headline in the San Francisco *Examiner*'s human interest section:

ALMOST HUMAN INTELLECT
An Astronomical Machine That Discovers Comets All By Itself
The Meteor Gets in Range, Electricity Does the Rest.

As Barnard read on, appalled, he learned about his own fictitious invention of a telescope that would search the sky by itself. It was a hoax, and whoever wrote the piece must have known what he or she was doing. It described how comet seeking is a difficult and time-consuming activity and how wonderfully easy this new machine would make it. On paper at least, the procedure was simple. The device would examine the spectrum of everything as Barnard slept—an unlikely act for him on a clear night. "Stars, nebulae and clusters innumerable crowd into the field [of view of the telescope] with every advance of the clock," the article went on, "but the telescope gives no sign of their presence. . . ." But should any object showing the "three bright hydrocarbon bands" of comet light appear, the device as described by the paper would allow that light to pass through and hit a selenium band, closing a circuit, and setting off an alarm in Barnard's bedroom. The sleepy astronomer would rush upstairs to the telescope, make a simple visual confirmation of the new comet, report it, and return to bed.[10] The clever idea of a selenium band being sensitive to light had been known for some 2 decades, and the public was becoming increasingly aware of its properties.[11]

Barnard became apoplectic. He wrote letter after angry letter, but somehow the *Examiner* had been warned that the reclusive scientist would doubtlessly deny everything, and the paper refused to print them. Although comet hunter Lewis Swift had his doubts that such a device would work, even he wrote to Barnard asking for all the details: "It takes my breath away," he gushed, "and makes my hair stand straight towards the zenith to think of it. Although the article appears somewhat fishy," he went on, "I am inclined to think it is still another of the marvelous inventions of the 19th century."[12]

Almost 2 years passed before Barnard was able to assure the *Examiner* that it—and he—had been had. Finally convinced, the paper apologized on February 5, 1893:

> The *Examiner* seizes the opportunity to express contrition for the annoyance which it caused this eminent scientist by printing some time ago an account of a highly ingenious, but non-

existent, machine for scanning the skies and catching wandering comets on the photographic plate.

At the end of the apology, the paper bequeathed him "all the new moons and comets that may be necessary to his happiness."[13] That same day, the paper ran an article on "How to Find Comets"—an authentic one this time, by Barnard himself.

Barnard never quite got over this hoax. As Heber Curtis, an astronomer and friend of Barnard's, wrote years later, "Even as he told me the story, ten years after the event, he was able to summon only a rather wan and rueful smile."[14] Nor was Barnard ever fully certain who perpetrated the hoax, although the *Examiner* admitted that Charles Hill, who had worked at the Lick Observatory in 1889, had something to do with it. The lack of any great effort by Lick Observatory to support Barnard in his denials led Barnard to believe that one of the senior staff members was behind the hoax. "Well, I never was sure," he told Curtis, "but I have always suspected Keeler."[15] Spectroscopist James Edward Keeler, according to Curtis, "was always ready to laugh at other astronomers, or at himself, if need be."[16]

Why would Keeler, who only 4 years later cofounded the prestigious *Astrophysical Journal*, do such an irresponsible thing and then let Barnard wallow in it for 2 years? One guess comes from John Lankford, who in 1978 wrote in *Sky and Telescope* that Keeler did it as a joke to shore up morale at Lick Observatory. Having worked in observatory environments, I certainly agree that something in the air—the altitude, long nights in a cold dome, or long years with the same few people—eventually creates so much tension that astronomers are at each other's throats. The atmosphere was poisoned at Lick at the end of 1890, with astronomers barely on speaking terms with one another.[17] Barnard himself left in 1895, when he was offered a position at Yerkes Observatory in Williams Bay, Wisconsin, where he stayed until the end of his life.

Curtis himself began his account with this delightful verse from Kipling:

> *The Doorkeepers of Zion,*
>   *They do not always stand*
> *In helmet and whole armour,*
>   *With halberds in their hand;*
> *But, being sure of Zion,*
>   *And all her mysteries,*
> *They rest awhile in Zion,*
> *Sit down and smile in Zion,*
> *Ay, even zest in Zion;*
>   *In Zion, at their ease.*[18]

Curtis didn't add that Kipling's humor grew out of a warning that appears in *Amos* 6:1: "Woe to them that are at ease in Zion." For while the hoaxers were at ease, Barnard put on his armour and unintentionally got the better of them. Eighteen months after the prank, in 1892, Barnard actually did discover a comet in a completely new way. Instead of using his eyes, he used the emulsion of a photographic plate attached to a telescope. On the plate, he found Periodic Comet Barnard 3 (1892 V) near the bright star Altair. Because the exposure was a long one, the moving comet appeared as a fuzzy trail extending from southwest to northeast. A second picture taken of the same area confirmed that the object was really a comet, and not a defect on the plate and that it was moving toward the northeast. It turned out to be the last of Barnard's 17 discoveries, but it started a whole new era of comet searching that continues to this day.

A Different Drummer

If one advances confidently in the direction of his dreams, and endeavors to live the life he has imagined, then he will meet with a success unexpected in common hours.

If a man does not keep pace with his companions, perhaps it is because he hears a different drummer. Let him step to the music which he hears, however measured and far away.

Henry David Thoreau[1]

*L*ike everyone else, comet hunters are a disparate group of people with diverse jobs and life-styles, but they never veer far from the direction of their dreams, a lonely patch of haze secretly lurking in the sky. The beat to which comet hunters keep pace is set by the moon, whose eastward march defines the dark hours for hunting. Because they hunt for a variety of reasons, some scientists have scorned their work; in 1930 for example, Harvard College Observatory Director Harlow Shapley praised and insulted Messier in a single sentence: "He is remembered for his catalogue; forgotten as the applause-seeking discoverer of comets."[2] And then only 4 years later, Shapley lauded comet hunter Leslie Peltier as "the world's greatest nonprofessional astronomer."[3]

THE PENDULUM SWINGS SOUTH

For most of us, Shapley is wrong about Messier. If comet hunters really seek applause for their play, they can wait years between acts. To find lots of comets takes a lifetime of work and dedication; to find more than one really bright comet takes luck. For all his skill, the Australian comet hunter John Tebbutt was incredibly lucky to find two comets—in 1861 and 1881—whose long tails would dominate the sky.

When Tebbutt discovered the great comet of 1861 from Windsor, New South Wales, on May 13, 1861—the one that made such a sudden grand entrance in the sky over England—it was a faint fuzzy spot, barely visible to the naked eye, and more than 6 weeks away from its best performance. Moreover since there was no reliable rapid communication south to north, the northern hemisphere was quite unaware that a bright comet was on its way. At the end of June, the huge comet's head was in the northern sky near the bright star Capella while its tail stretched more than halfway across the heavens to the constellation of Hercules. Apparently the spectacle became even better than that, for the Earth went right through the outer reaches of the comet's tail on the night of June 30.

Unlike many of today's comet hunters, Tebbutt made a point of observing comets as well as discovering them. In fact after discovering the comet of 1861, he calculated its first preliminary orbit. Since he kept careful journals of all his observations, we are able to read about this remarkable observation:

> In the evening of June 30 I observed a peculiar whitish light throughout the sky, but more particularly along the eastern horizon. This could not have proceeded from the moon, but was probably caused by the diffused light of the comet's tail, which we are very near to just now.[4]

By 1863 Tebbutt had completed his own small observatory by himself, a structure that would last for many years and include a transit instrument for accurately measuring the positions of comets. From his new observatory, he independently discovered Tempel's comet around the same time that Tuttle and Hopkins were observing it from opposite shores in August 1864.

In 1881 Tebbutt found a second great comet. On the evening of May 22, without a telescope, Tebbutt spotted a faint fuzzy object in the constellation of Columba the dove. "Immediately on its discovery I obtained," he wrote, "with the 4.5-in. equatorial, eight good measures of the nucleus from one of the bright stars just mentioned. On the following day I notified the discovery to the Government Observatories of Sydney and Melbourne."[5] By the end of June, the comet had brightened to first magnitude. Considering how few observers were in the southern hemisphere in the nineteenth century, Tebbutt did a big favor to history by making his comet observations as complete as possible. When he retired from active work in 1904, he had completed exactly 50 years of comet observations.

THE PENDULUM SWINGS EAST

Despite the poor weather in England, William F. Denning was another great comet hunter who managed to compete successfully on the comet playing field. His greatest contribution however came in observing meteors. Denning was passionately involved with observing these pieces of space rubble that are the remnants of passing comets. Sitting in an eastward-facing chair on warm nights, standing up and moving to keep warm on cold nights, he recorded more than 9000 meteor trails over a 17-year period, coming from no fewer than 918 different points in the sky, called radiants. Because of the effect of perspective, each meteor shower has its own radiant. Although many of Denning's radiants are no longer recognized, his sheer volume of observational work is astounding. His observations showed, for example, that the Perseid radiant moved from night to night as its aspect with relation to the Earth changed.

On the morning of July 11, 1881, Denning planned to hunt for comets in the pentagon-shaped figure of Auriga, which was just rising in the northeast before dawn. But morning observing can be quite a challenge for tired eyes, and he decided to forego his sweep. On July 14, from Ann Arbor, Michigan, John Schaeberle did hunt through the pentagon and picked up a bright sixth-magnitude comet there. Comet Schaeberle 1881 IV was a tough blow for Denning, for it grew a long tail and brightened to third magnitude—easily visible to the unaided eye—by the end of August. Even under normal conditions, a fairly bright comet in the north circumpolar summer sky would have really been a special treat. But Tebbutt's major discovery, the now-fading great comet of 1881, was only 8 degrees from the pole at the end of July, and for a few weeks, two naked-eye comets were in the same part of the sky at the same time.

Denning had a chance to recover from his disappointment only 3 months later when he did not pass up a chance to sweep the morning sky in Leo on October 4 and found the periodic comet now known as Denning–Fujikawa. (Under today's rules, Brooks would have his name attached to the comet as well, as its third independent discoverer, but at the time, the honor of having your name attached to a comet belonged to only the first one to find it.) Two weeks later Denning suggested that English observers could easily join the

> friendly competition with foreigners. . . . We might act in unison, and arrange a division of labour. Certain regions could be apportioned to different observers; in any case it is very desirable to adopt some method likely to command the interest and labour of observers, and to give a fair measure of success.[6]

The idea of having comet hunters join in a group and divide the sky is good in theory but has rarely succeeded in practice. Different locations have different weather patterns, and the only procedure that has really worked is to have comet hunters sweep the sky from several locations. Moreover comet hunters by nature bridle at the thought of restrictions: They will search through as much sky as possible on a dark, clear night.

At the turn of the century, William Reid, another British amateur astronomer, moved to South Africa, where he discovered Comet Reid 1918 II in June 1918. Within the next 8 years, he found an additional seven comets. Not only did Reid dislike calling attention to himself, but he was also notably generous toward his fellow comet hunters. As a story goes, he once found a comet (probably 1926 III), but before he reported it, he heard that G. E. Ensor, an acquaintance, had independently picked it up as well. Realizing that his competitor had never discovered a comet and feeling that he had enough comets to his credit already, Reid declined to report it himself. Thus the comet Reid is known as Comet Ensor 1926 III.

This comet disappeared for a few weeks as it approached perihelion, and it was too close to the sun to be seen. But only 11 days after the comet rounded the sun, a group of Russians floated into the sky beneath a tethered hot-air balloon. As the balloon climbed past 2000 feet, S. M. Selivanov, a member of the Mirovedenie Society that sponsored the event, spotted the comet through a pair of binoculars.[7]

BACK TO THE STATES

In the United States, three promising comet seekers made their first finds in the first decade of the twentieth century; one was Zaccheus Daniel, who used a 6-inch diameter refractor from the Princeton University Observatory to discover three comets, one of which (Comet Daniel 1907 IV) reached second magnitude and had an easily visible tail. Joel Metcalf of South Hero, Vermont, discovered comets in 1910, 1913, and two in 1919. The third comet seeker was John E. Mellish, a telescope maker who observed from Madison, Wisconsin. He found two comets in 1907 and two more in 1915. But then the story takes a different direction. In 1915 Mellish responded to an ad that Jessie Wood, of Glencoe, Illinois, placed in Chicago newspapers:

Wanted—a perfect husband, one who wants that happiness
not of a day, but of a lifetime; who would receive the fullest
pleasure in staying home at night talking to me and would
be just as wrapped up in me as in his work.

Out of some 2000 young men who responded to the ad, Jessie
chose Mellish, and they were married that year. One wonders if
she were expecting someone who would want to stay up all night
searching the sky for comets. In 1917 Mellish found a fifth comet,
in 1923 he bagged a sixth, and in 1932 he was in Geneva County
jail. Apparently he hadn't turned out to be the perfect husband.
His wife left him and accused him of chasing after a 15-year-old
girl. Although he admitted to the charge, he entered a plea of not
guilty. According to a 1932 newspaper article,

Prosecutor Carbary and Circuit Judge John K. Newhall, under
whose jurisdiction Mellish is, admitted today they were in a
quandary over disposition of the case. They said a dozen letters
from scientists of seven universities had appealed for mercy.

This episode marked the only interruption in a career that was
devoted first to comets and later to telescope making. Mellish then
moved to California, where he resumed his career and made telescopes almost until his death in 1970.

WALDEN OF THE SKY

Two years after Mellish's last comet discovery in 1923, on a
farm near the small town of Delphos, Ohio, Leslie Peltier, born
on January 2, 1900, made his first discovery. Peltier's interest in
astronomy was sired by sightings of two major comets in 1910,
the Great January Comet, and Halley's, but the real spark came
on a May evening in 1915: "Something—perhaps a meteor—
caused me to look up for a moment. Then, literally out of that
clear sky, I suddenly asked myself, 'Why do I not know a single
one of those stars?' "[8]

To raise the 18 dollars needed to buy his first telescope, Peltier picked 900 quarts of strawberries on the family farm at 2 cents a quart. Three years later, he started observing stars whose brightnesses change with time and sometimes unpredictably. He was thrilled with the antics of these variable stars. On March 1, 1918 he made the first of what would become a lifetime total of 132,000 variable-star observations.

As Peltier's reputation as a skilled observer spread, he was offered the use of a 6-inch refracting telescope by Henry Norris Russell of the Princeton Observatory. Peltier learned that Zaccheus Daniel had used this telescope to discover three comets, including the impressive naked-eye one in 1907. Knowing that his new telescope was a proven comet finder gave Peltier a connection to a tradition. "It seemed to me," Peltier wrote, "that if ever human attributes would be invested in a thing of metal, wood, and glass, then this ancient instrument now in my keeping must long for one more chance to show what it could do."[9]

It was 3 years before Peltier tasted success in his comet searching. Starting at the horizon, he described the search through his opened observatory dome:

> I slowly worked upward back and forth in horizontal sweeps across that bounded bit of sky. Sweep by sweep I climbed upward through Corona, pausing ever so briefly as I hailed, in passing, the patterned landmark of the R Coronae field, and on I moved into the northern end of Bootes.
>
> It was just above the peak of that kite-shaped figure of the Herdsman that the steady cross-sweep of my telescope abruptly stopped. A small, round, fuzzy something was in the center of that sea of stars! A closer, calmer look and I was sure just what that something was, for extending downward from it I could dimly see a slender streak that could only be the tail of the comet I had just discovered![10]

The comet was moving so rapidly southward that initial attempts to confirm it were unsuccessful. But A. Wilk, a high school teacher in Krakow, Poland, independently discovered the comet, and thus it was saved. According to Wieslaw Wisniewski, on December 20, 1929 Wilk found his second comet and his third just 3 months later. In 1937 Wilk and Peltier discovered their second comet together, but the comet was announced in Europe as Comet Wilk before Peltier could report it. This was Wilk's final comet discovery. The Nazis incarcerated him along with other teachers and university professors during the war. After heavy international protests, some of the prisoners, including Wilk, were released. But it was too late for the comet hunter, who died shortly after.

FROM FLYING SANDBANKS TO DIRTY SNOWBALLS

By midcentury, astronomy's understanding of the nature of comets was about to take its biggest leap since Halley's breakthrough more than two centuries earlier. As a graduate student at the University of California at Berkeley in 1930, Fred Lawrence Whipple was part of a team that calculated a rough orbit of the newly discovered planet Pluto. By the end of World War II, Whipple, who was now at Harvard, noticed that the orbits of meteors, which are very much like those of comets, were slowing down, and that the meteors should have all fallen to the sun. Something was replenishing the solar system's supply of meteors. Comets were then thought to be huge flying sandbanks. Whipple decided that the sandbank model could not provide a current source for the meteors; instead he proposed that comets were large conglomerates of ices and meteoric particles, now popularly known as dirty snowballs.[11] By September 1951 Fred Whipple had published the results of his work on the orbit of Encke's comet in two of the most important papers in the history of comet science.[12] After four decades of observations, Whipple's comet model was about to be dramatically confirmed when the European *Giotto* spacecraft passed by the nucleus of Halley's comet in March 1986. "Well, Fred," project scientist Rüdeger Reinhard declared, "tomorrow is your moment of truth."

"Yes," the 79-year-old Whipple said. "Things are going to get truthier and truthier."[13] *Giotto* sent back beautiful pictures of a potato-shaped comet nucleus that reaffirmed Whipple's grand reputation as Mr. Comet.

GREAT COMET OF THE SIXTIES

By 1945 comet hunting had become worldwide again. A team of Czechoslovakian observers, including Antonin Mrkos and Ludmilla Pajdusakova, using large 25 × 100 binoculars, found the first of 16 comets at the astonishing rate of 100 searching hours per discovery—about half of what most other observers typically spend.

On August 2, 1957 Mrkos saw the tail of a bright comet rising, and the new surprise visitor was quickly announced as Comet Mrkos. A short time later, Walter Scott Houston, a *Sky and Telescope* columnist, was enjoying his pipe after dinner when Clifford Simpson asked why he had not pointed out a bright comet in the west. Assuming that Simpson was speaking of Comet Arend–Roland, which had been bright some weeks before, Houston, not missing a puff, said that there was no comet. "But there is a comet there!" Simpson protested. Finally Houston gave in, looked up and saw one of the brightest comets of his life. He darted to his telescope, yanked it with one hand, focused it with the other, stashed the lit pipe in his pocket, and observed Comet Mrkos.

"Scotty, you're on fire!" Simpson hollered. Ripping off his burning jacket and throwing it on the grass, Houston was about to return to his telescope when the grass also caught fire. Finally they smothered the flames just in time to lose the comet behind a tree.

By the mid-1950s, Minoru Honda was finding comets from Japan. His record of a dozen comets inspired a generation of Japanese comet hunters, of which the best known were Kaoru Ikeya and Tsutomu Seki.

I first heard about Ikeya when he discovered his first comet early in 1963. Ikeya had felt that his family's name had been dishonored because of his father's business failure and subsequent drinking problem. Leaving school to take a job at a piano factory, Ikeya built an 8-inch telescope for the equivalent of 20 US dollars, and in 1961, he began regular comet searches. Although New Year's Eve 1963 was clear, his family asked him not to search that night and instead to celebrate the holiday. But just before dawn on the following night, Ikeya found a comet in Hydra. Brightening rapidly as it approached the sun, the comet reached third magnitude by March 21. With a comet named after him, Ikeya believed that his family's name had been rehabilitated throughout Japan. And a year later he found a second comet.

Tsutomu Seki, a guitar instructor, was also building a noble record of discoveries by 1965, including a comet that reached naked-eye brightness in 1962. Seki's 1962 find was awaiting confirmation when Richard and Helen Lines, a couple observing near Phoenix, Arizona, also snatched it.

On September 18, 1965, Ikeya was again sweeping in the southern constellation of Hydra when he found an eighth-magnitude comet. Creeping southeastward, the comet was an exciting find for Ikeya, who was now bagging them at the comfortable rate of one per year. But Seki was comet hunting that morning too, and 15 minutes later, he found the same object. Thus confirmed, Comet Ikeya–Seki 1965f was announced to the world by Owen Gingerich, then director of the Central Bureau for Astronomical Telegrams.

This comet's motion was not immediately obvious; in fact the first orbit, issued on September 28 and based on just 2 days of observations, had the comet making a leisurely pass around the sun at slightly less than the distance of the Earth.[14] But with new, accurate positions, in the last week of September Fred Whipple recognized that Comet Ikeya–Seki was extraordinary: "The following parabolic orbit," the International Astronomical Union *Circular* announced, "shows a close resemblance to that of the Great sun-grazing Comet of 1882 and its family. According to B. G.

Marsden, Comet Ikeya–Seki may be as bright as −7 [much brighter than Venus] at perihelion."[15] The comet of 1882, discussed earlier, did a hairpin turn around the sun. At eighth magnitude a month before perihelion, Ikeya–Seki was coming around at the same time of year as the 1882 comet; the Earth was in its most favorable position for a near repeat of the fabled show.

But time was running out. With 3 weeks until perihelion, there was barely enough time to inform everyone that something really big was coming in. To speed things up, on October 1 Gingerich and Marsden, who had just finished his thesis at Yale and was in his third day of work at the Smithsonian Astrophysical Observatory at Harvard, held a press conference to announce that a probable great comet was on its way. Circulars were released frequently with updated orbital elements and visual magnitudes showing that the comet was brightening just as was expected.

Sun-grazing comets are exceptional because they whip around the sun at less than a million miles from its fiery surface. First studied in detail as a single group with a possible common origin by Heinrich Kreutz in 1888, these comets share a similar orbit. Their grandeur depends on the time of year they round the sun. Comet Pereyra, at perihelion in 1963, was bright but far from spectacular. Both the 1882 and the 1965 comets rounded the sun not far from the September equinox, squeaking past the sun at less than 300,000 miles and putting on splendid shows. According to Marsden, these two comets almost certainly last brushed the sun as a single object, perhaps the great comet of 1106.[16]

As the comet approached its October 21 date with the sun, everyone with the slightest sense about the sky was excited. Would the comet immolate itself as it tried to touch the solar surface? Hoping to get a view before perihelion, I set my alarm for predawn every morning, but clouds always blocked the sky. With 3 days to go, Sunday morning, October 17, was clear. I biked a couple of miles from my Montreal home to Summit Park, a site with a good view to the east, which I thought would be the best place. So did a hundred other people: at 5:00 A.M. the park was jammed. But although most of the sky was clear, clouds low in the southeast prevented us from seeing the comet.

Perihelion day was October 21, 1965. From the clear locations in the southwestern United States, the comet was clearly visible to the naked eye at magnitude minus 8, virtually the brightness of a quarter moon. By the twenty-fourth the comet, now visible in a dark sky, had a 20-degree tail.[17] But in eastern Canada, I was still waiting for a clear morning. When I awoke on Friday, October 29, the clear sky outside my window sent a chill through me. I dashed downstairs and hopped on my bike, racing toward Summit Park in time to beat the dawn. Completely out of breath, I reached the Summit lookout and stopped. The view that morning was splendid. Beyond the expanse of the city lights, beyond the distant St. Lawrence River, a mighty searchlight beam rose out of the southeast like an exclamation mark. The crowd from the week before was gone; it was just me, the sky, and my first view of a comet.

Two mornings later and across the world, Bart Bok, director of Australia's Mount Stromlo Observatory, awoke from his sleep. His attention was sparked by what he, like me, thought was a bright searchlight beam dominating the eastern sky. Astonished by the sight, he quickly grasped the opportunity to show the public this majestic apparition. Within the next few hours, the media all over Australia had published Bok's account of the comet, and on the morning of November 1, thousands of Australians arose to see it.[18]

By November 6, several observers reported that the nucleus had split into at least two parts. Later it was predicted that the larger one would possibly return in 8 centuries, the smaller one 3 centuries later.[19]

STARLIGHT NIGHTS

If this wonderful comet was not enough to excite my interest, Peltier's autobiography *Starlight Nights* appeared the same year. The book was a sky gazer's *Walden*, clearly a call for comet hunters.

A decade later, I visited Peltier. In a conversation about comets, variable stars, and politics, he was as vibrant as he had been in his writing. He had found his twelfth comet in 1954, and since a large 12-inch refractor had come his way, he had devoted much of the rest of his observing life to variable stars. "I don't think amateurs have it so easy any more in comet hunting," he said. "Professional photographic searches make it quite difficult for an amateur to come in first." Should we give up then? I asked about today's searchers. "Of course not. You just have to try harder."

During that first visit, I realized how right Shapley had been about Peltier's greatness but how utterly wrong he had been to generalize about Messier's seeking applause for his comet finds. Here was a man so modest that he rarely went to astronomy club meetings. Peltier's friend Walter Scott Houston says that once, to get him to an important variable-star observers meeting, some members virtually had to kidnap him. When presented a major award from the Western Amateur Astronomers in 1967, Peltier had to be strongly encouraged to leave his Delphos, Ohio, home to go to the West Coast. Not much of a traveler, he particularly disliked flying. "Why did you take the train to get all the way out here," someone asked Peltier at that gathering. "I had no choice," he explained. "The stage coach no longer runs."[20]

In fall 1979, I met Peltier for the last time. It was during this visit that I realized how much Peltier's home and telescopes were a part of him. He showed me a corner of the transit room of his 12-inch Clark observatory, where the old mahogany tube sat that at an earlier time had held the lenses of his 6-inch comet seeker. There, carved in deep, proud letters that almost encircled the little tube, were the designations of the three long-gone comets of Zaccheus Daniel and the 12 Peltier comets whose light had passed through that telescope before any other. These hand-cut numbers and letters were the core of a career.

Fifteen times this little telescope had uncovered a new member of our solar system, and the unassuming person standing next to it was responsible for 12 of them. What Peltier did could profitably be done by amateurs today; what he achieved is a worthwhile goal to which any modern observer can aspire. And the mood, the attitudes, and the way of thinking he brought to our avocation have given us a special pride in what we do.

More than any other comet seeker before him, Peltier understood what Thoreau's different drummer really meant to a comet hunter. "Time has not lessened the age-old allure of the comets," he wrote in *Starlight Nights*.

> In some ways their mystery has only deepened with the years. At each return a comet brings with it the questions which were asked when it was here before, and as it rounds the sun and backs away toward the long, slow night of its aphelion, it leaves behind with us those questions, still unanswered.

> To hunt a speck of moving haze may seem a strange pursuit, but even though we fail the search is still rewarding, for in no better way can we come face to face, night after night, with such a wealth of riches as old Croesus never dreamed of.[21]

The Comet Cop

*W*hen bright Comet Mrkos suddenly appeared in the western sky in August 1957, a young summer student named Brian Marsden was working in his basement office at the Royal Greenwich Observatory, then located at Herstmonceux Castle in southern England. He was not interested in discovering comets; his passion was calculating their orbits as soon as possible after they had been discovered. And over the next 35 years, Marsden's knowledge of where every comet was at virtually any moment made him one of the most important figures in the comet field.

But during the summer of 1957, Marsden was just an undergraduate student trying to calculate an orbit for a comet on an old mechanical calculator. He had two accurate positions for the comet, but he needed at least one more to calculate the comet's orbit. One evening the telephone rang. At the other end of the line was a voice with the numbers for that badly needed third position. Marsden excitedly sat down to try his hand at the orbit. But it didn't work; no matter what mathematical tricks he used, there was no way that the comet's first two positions would allow it to reach the third one. He could not obtain any kind of an orbit from those positions. "It took me more than an hour to conclude that this observation must have been a hoax," said Marsden, ruefully recalling his wasted evening.

One of the brightest comets of this century, Comet Mrkos's sudden appearance in the evening sky was electrifying for every sky watcher the summer of 1957, and Marsden had an unexpected opportunity to observe it. An evening or two after he had received the bogus comet position telephone call, two young female students warned Marsden of a plot to kidnap him. He was to be handcuffed and driven to parts unknown, the women said, and then he would have to hitchhike back. "I like to think the perpetrators were concerned I was working too hard," Marsden says, a charge many of his colleagues have against him to this day.

So with his newfound allies, Marsden resolved to leave his office discretely that evening, and the three sat on the hillside overlooking the castle. Low in the west, Comet Mrkos dominated the warm summer evening, its long dust tail streaking through the sky. But then they heard some commotion from down near the castle. Unable to find Marsden, the kidnappers' plot was unraveling and one of the kidnappers had become the kidnappee! Hours later the hapless victim returned. Enraged, he went to the other perpetrator—now the distinguished director of a well-known observatory—and threw him into the castle moat.

Two years after the Comet Mrkos episode, Marsden came to the United States, exchanged his mechanical calculator for a klunking IBM 650 computer, and eventually got his doctorate at Yale. On the day after Christmas 1964, he married Nancy Lou Zissell. Their daughter Cynthia remembers, at age 4, when spectacular Comet Bennett lit up the morning sky in 1970—she referred to it at the time as "skywriting."

Marsden arrived at the Smithsonian Astrophysical Observatory some 10 days after two Japanese amateur astronomers, Kaoru Ikeya and Tsutomu Seki, discovered their new comet in September 1965. On Marsden's first day in Cambridge, he was put to work refining the orbit of what was going to be the best comet to graze the sun since 1882. World-renowned astronomer Fred Whipple had just suggested that this one might be a member of the celebrated group of comets described by turn-of-the-century German astronomer Heinrich Kreutz. With 3 weeks until perihelion, there was barely enough time to inform the public. On October 1 Marsden and Owen Gingerich, who at that time directed the Central Bureau for Astronomical Telegrams, held a press conference to announce that the bright comet was on its way.

Since those Ikeya–Seki days, the Kreutz sun grazers have been one of Marsden's particular interests, and he wrote an extensive paper on them in 1967.[1] His article includes an ephemeris for sun grazing comets at specific distances from the sun. Such a table is especially useful for those seeking new sun-grazing comets.

WHO DISCOVERS WHAT

Marsden succeeded Gingerich as director of CBAT in 1968. Marsden now decides who receives credit for finding anything astronomical that moves or explodes; perhaps he is more accurately called a policeman of the sky. Marsden knows where the comets and asteroids are in their orbits, and he makes sure that the 50- or so exploding stars—supernovae and novae—that appear each year are announced promptly.

More than anybody else on Earth, Marsden knows where things are in the sky. One key to his remarkable success comes from his ability as a celestial mechanician; he has been calculating orbits since high school. He also has a phenomenal memory: He doesn't forget details of events or their place in time. One afternoon I showed him my backyard observatory with its decade-old dedication sign: November 15, 1980. "Ah, yes!" Marsden laughed. "That was a Saturday."

Relying more on postcard circulars and electronic mail than on telegrams, Marsden and associate director Daniel Green, who has been at CBAT since 1978, are now issuing circulars at the rate of about five a week. Green brought to his job the active interest in comets that he had had since high school, when he edited the *Comet*. In the last decade, Green's journal, now called the *Inter-*

national Comet Quarterly, has been read widely by comet enthusiasts around the world.

The idea of CBAT circulars is to report events in the sky efficiently, although in 1970 an earthly shooting found its way into *Circular* 2209. The event took place in West Texas at McDonald Observatory's new 107-inch telescope. "Shortly before CST midnight Feb. 5," according to the story,

> a newly-hired employee fired seven point-blank shots into the front surface of the fused-silica primary mirror, using a 9-mm pistol. . . . The harm suffered by the mirror from his bullets, and his several preliminary blows with a hammer, was extraordinarily small.

At the end of August 1975, a bright nova in Cygnus lit up the sky. Although Marsden quickly relayed word to the astronomical community by telegram, dozens of independent discoveries of the stellar explosion kept his telephone ringing at home and at the office all through the long Labor Day weekend. The circular he was finally able to send out after the holiday showed that a panorama of discovery reports had been made round the world as the Earth whirled eastward and observers from country to country looked up at the brightest new star in a generation. The brightest nova was followed by the brightest comet just a few months later. Comet West grew significantly more impressive than expected, its brilliant tail lighting up the morning sky.

In 1978 Marsden took over the directorship of the International Astronomical Union's Minor Planet Center, which moved from Cincinnati to Cambridge together with astronomer Conrad Bardwell as its associate director. On Bardwell's retirement in 1990, his place was taken by Gareth Williams. Since 1987 Marsden has also served as associate director of the Planetary Sciences Division of the Harvard–Smithsonian Center for Astrophysics, one of the nation's top schools.

During all this time, Marsden has also tried to maintain an active research program, including in particular his extensive pioneer work on nongravitational forces—the jets and other cometary eruptions—that affect the motion of comets like Halley's. Understandably Marsden has generally tailored his research papers to his activities as director of the central bureau. Marsden's main interests have been in methods for determining comet orbits based on a few early positions, and in the large spherical shell of comets at the edge of the solar system, called the Oort cloud after the late Dutch astronomer Jan Oort. Marsden has even found time to bring out eight editions of his well-known *Catalogue of Cometary Orbits* and a catalog of minor planet discoveries.[2]

Over the years Marsden has become an inspiration to would-be discoverers and other observers, pushing them to do more when he thought they were getting a bit lazy and cheering them up when they became discouraged. On one depressing day, I wrote him a note saying that perhaps I have found enough comets. "We have plenty of room for many more comets Levy," he replied, "so get out there and keep looking!" In 1983 George Alcock, who had visually discovered several comets *and* novae, complained that his discovery career was likely to end as organized professional searches took over. Marsden's reply typifies how he pushes amateur astronomers to keep looking, and keep succeeding:

> The funny thing is that in this modern mass-produced world so many groups and organizations set themselves up to make astronomical discoveries. Here is this group going to pick up supernovae in other galaxies by the hundred. That automatic telescope in Arizona is going to sweep up all the comets that come reasonably far above the horizon. This infrared-astronomy satellite is going to find scores of earth-approaching asteroids way ahead of Palomar. But I'm still waiting for these discoveries. Their participants have talked endlessly about what they can do—but they have produced absolutely nothing! You—and a handful of other dedicated individuals around the world—go out, without a lot of fanfare, and you make the discoveries. Sometimes years may pass between discoveries (occasionally only five days!), but they do come, because you know what you are doing, although you never know whether chance is going to be a friend or an enemy. So I'm ready now for your next discovery![3]

Comet IRAS–Araki–Alcock 1983d. (Photograph by Jack Newton, May 9, 1983.) This comet was moving so fast that Newton had to move both axes of the telescope during the 6 minutes the camera shutter was open.

Only 3 weeks later, astronomers working with the new infrared astronomical satellite (*IRAS*) thought they had discovered a new near-earth asteroid during the satellite's survey of the sky. Meanwhile Alcock knelt in his upstairs hall, directed his trusty binoculars through the double-glazed windows, and discovered (along with an independent amateur Japanese astronomer) what came to be known as Comet IRAS–Araki–Alcock 1983d, the comet that made the closest known approach to the Earth since 1770.

FALSE ALARMS AND HOAXES

Determining the priority in a discovery is one of Marsden's most sensitive tasks. Since comets started bearing the names of their discoverers in the late eighteenth century, their discovery has been comparable to finding the Holy Grail. I spoke with someone in Brussels who said that he "would do anything—absolutely anything—" to get his name on a comet. I wondered, though, if anything included what one really needed to do—spend hundreds of hours with his or her eye glued to an eyepiece at all hours of the night.

In his quarter-century at the helm of the CBAT, Marsden has received few phone calls in the middle of the night, "and never a useful one," he adds. About 90 percent of reports of new comets turn out to be false—mostly ghost images of bright stars or planets just outside the telescope field, photographic flaws, or two different galaxies thought to be a moving comet. Occasionally there are outright hoaxes: Once when someone from Columbus, Ohio, reported a comet, Marsden replied that he would wait for confirmation before issuing a circular. Some days later, a telegram arrived under the guise of the Tokyo Observatory with the required confirmation. However the telegram's return address was not Tokyo but Columbus, Ohio! Another time an observer tried to claim credit for a comet that had already been announced; he alleged to have observed it the previous night. This time Marsden noted that the would-be discoverer could not have observed the comet, since it had been raining all that night at the site.

Other false alarms have been easier to identify. One didn't even involve Marsden's office, at least at first: A young observer in Western New York announced that he had discovered a comet—but only to the local press and his astronomical society. In the following days, both the newspaper and the society were hoodwinked into publishing follow-up observations. Whether the erstwhile discoverer was deliberately lying or his moving comet was simply a series of observations of different galaxies is still uncertain.

Marsden and Green have amassed a great deal of experience in the field of comets as well as with those who observe them. Since so many reports turn out to be false alarms, the two do not routinely investigate all comet discoveries; instead it is up to a discoverer to confirm the existence of a particular comet. If the new comet is bright enough, another discoverer will probably find it quickly, and the comet will bear the names of both finders. Until a comet is confirmed, up to three discoverers, or two at the same observatory, can have their names attached to it.

Marsden has dealt with a host of other suspect sightings, including one luckless observer who sends his many reports of nonexistent comets by Federal Express, and a would-be-discoverer in Canada who reports the positions of new comets on the basis of his dreams! As we noted earlier, Barnard is the only comet hunter I know of who actually did discover comets after dreaming about them. Marsden generally treats these strange cases with a good deal of patience and even laughs about them when he can. I know of one person who tested his patience to the point that Marsden finally called him an ignoramus. But when that individual showed signs of mending his ways, Marsden immediately worked with him and encouraged his new attitude.

A more serious episode began in July 1980 when an amateur astronomer in Florida sent CBAT a telegram about an object of magnitude minus 25, almost as bright as the sun. Dismissing it as a hoax, Marsden went away for a few days on a rare vacation. Daniel Green then received a phone call from the Florida observer, who was angrily wondering why nothing was being done about the asteroid he had discovered near the sun and could be seen in the glare as it moved off the sun's limb. The amateur was not convinced by Green's logical suggestion that the solar obscurer was probably a weather balloon launched from a nearby air force base. On his return from vacation, Marsden was subjected to a 90-minute Saturday afternoon tirade from the Floridian, who by this time was trying to persuade US government officials of the need for defense against the killer asteroid that he (and apparently no one else) had seen.

The persistent amateur also contacted Tom Van Flandern, then on the staff of the U.S. Naval Observatory, who suggested some 30 other possible explanations for the observation. This time the Florida observer responded with letters to higher-ups in the Naval and Smithsonian managements. Then one day in August 1981, the Florida observer suddenly entered Marsden's office. The subsequent argument became so loud that people down the corridor rushed out of their offices. "Do you want a bouncer?" someone asked. Surrounded by three or four burly astronomers, the unfortunate character rushed out of the building. The sad episode didn't end there, however; for several years afterward, the Florida observer continued to send abusive and even threatening postcards to Marsden, Van Flandern, and Green. Finally the Florida observer committed suicide.

JUST A COMET SUSPECT

All these incidents are part of Marsden's life. My own experience with him includes my reports of 19 new comets, and his request to confirm the discoveries of several others. Confirming other people's finds should be part of the game for every searcher. One December night I was searching with my 16-inch telescope, moving it back and forth through the western sky. Then the telephone rang. It was Doug George, an experienced amateur astronomer from Ottawa, Canada. He had just found a new comet, also in the western sky, and wanted to know if I could confirm its existence for him. When he gave me the comet's position I did a doubletake—if I hadn't answered the phone, I might have discovered it myself within 10 minutes! Another time Marsden asked me to confirm a comet found by Mauro Zanotta in Italy. Although night had begun there, it was still early in the afternoon when I learned of the discovery. That evening I set up my telescope under a partly cloudy sky, turned it to the position Marsden had provided, and there was the new comet. When Marsden learned of my observation, he was about to announce the comet as Comet Zanotta. But then Howard Brewington independently located it from his home in New Mexico, and the comet's name became Zanotta–Brewington.

I remember fondly a day in 1987 that began when I sent Marsden by electronic mail a routine report of a magnitude estimate of Comet Bradfield, which had been prominent in the evening sky that fall. What wasn't routine was that my report ended with a note that I had spotted a comet suspect in the constellation Bootes. Marsden answered immediately, wishing me luck that the new comet was indeed a real one. In the next several hours, we exchanged perhaps half a dozen messages. Could I take a look at a possible comet reported by someone at such-and-such a position? Yes I would, I promised. Then he discussed a man who had bothered both of us as well as almost every other astronomer in Arizona and his theory that planet X may be somewhere, as yet unreported. Supported by a series of biblical passages, this man was convinced that the new planet was now visible. He had shown me a picture of it—it looked like streetlights in a parking lot.

As the afternoon wore on, I got a message from Canadian astronomer Terence Dickinson that he had confirmed my comet discovery from his home near Kingston, Ontario. I sent that along to Marsden, and suddenly the relaxed tone of our electronic mail was over. "Please send the positions from Dickinson as soon as possible, so the announcement can be made." I was typing the all-important Dickinson position as well as my own second night's observation when the phone rang. It was my planet-X friend. "We think we have planet-X," he said, interrupting my concentration. "What is its position?" I asked tersely. "Oh, I won't reveal that information," he replied. The following day all was back to normal. The new comet was announced as Comet Levy 1987y, the suspect Marsden asked me to check turned out to be nothing, and the planet-X guy was mailing his parking lot lights picture to Marsden.

As part of his job's responsibility, Marsden is plugged in to the astronomical community—both the scientific aspect and those who participate in it—more than anyone else. For the many observers around the world who rely on CBAT for announcements and orbits, Marsden has become as much a part of the celestial landscape as the objects they report.

☾ 8 ☽

Comet Tales

Comet hunting has attracted the fancies of many, including
William Brooks, who, in the late 19th century, hunted in his
yard with a nine-inch refractor and picked up over twenty
comets, Charles Messier, better known for his "non-comets,"
Leslie C. Peltier, who between 1925 and 1954 gathered twelve
comets and an assortment of novae, and David H. Levy, who
between 1965 and 1970 has found nothing—absolutely
nothing.[1]

*In fall 1965 I enjoyed watching Comet Ikeya–Seki paint a beautiful
picture as it did a hairpin turn around the sun. I decided then that it*
was time to begin a comet-seeking program of my own. I felt ready
for it. Over the past 5 years, I'd trained myself in three areas of
observing that are basic to beginning a search for comets. Incon-
gruous as it might appear, observing the moon in detail is one of
them. When I first walked into the Royal Astronomical Society of
Canada's Montreal Centre in 1960, I was presented with a simple
map of the moon on which 300 craters and 26 mountain ranges
and other features were plotted. The idea was to identify each of
these, an exercise that helped train my eyes to concentrate on details
near the limit of vision. Observing the soft markings on Jupiter,
Mars, and Saturn helped the process along.

In August 1962 a single observation of the Pleiades, Messier 45, began my second stage of observer training. Observing and describing all 110 objects in Messier's catalog is a superb way of becoming familiar with the sky and all the different types of fuzzy objects that it holds.[2] We know that Messier used a small telescope for most of his objects and a modern telescope would not easily mistake brighter objects for comets. But other objects, like Messier 78 in Orion, can still fool the unwary. Though Messier 78 is a nebula shining by the light of newborn stars more than a thousand light years away, it actually looks a bit like a tiny nearby comet. Although it took me 5 years to finish the list—that happened in May 1967—the second time around went a bit faster. On a single clear night in March 1983, I found, observed, and took notes on all of the Messier objects except Messier 30, which rose after the dawn was too bright.

The sky is a large place, and there are plenty of people moving through it with their telescopes in search of comets. At the end of 1967, Ikeya and Seki shared a second comet, and the following year, Minoru Honda bagged a nice one. One night while hunting outside my parents' home in the middle of a light-polluted city under an almost-full moon, I made an independent discovery of Comet Honda. Considering the poor sky, it still appeared quite bright through the 8-inch reflecting telescope I was using. I realized it had to be a known comet, and it did not take long to find out that it was Honda's comet. I remember my grandmother was visiting at the time, sharing my excitement as I darted in and out of the house between my star charts and my telescope.

SEARCHING FOR A DARK SKY AND A COMET

The next year, Comet Tago–Sato–Kosaka 1969 IX interrupted my college studies at Acadia University in Wolfville, Nova Scotia. This was the first comet to be observed by a spacecraft, the *Orbiting Geophysical Observatory*, which detected a huge envelope of hydrogen gas around the coma. And a few months after this, bright Comet Bennett graced the morning sky in March 1970.

That fall I set up my telescope for hunting at a site a mile north of campus. I must have had a lot of energy then, because I actually walked that mile with telescope, mount, and notebook in hand to set up for an evening's session. But this night I set up the tripod, put the telescope on the tripod, put an eyepiece in the telescope, and looked through it. Near the top edge of the field was a comet! It turned out to be a known one, a comet that Japanese amateur Osamu Abe had snatched earlier. Would that my first new one be that easy!

During the 1970s my comet hunting dropped in intensity thanks to my heavy course load, and by 1979, I still hadn't found anything. I began to think about what I could do to increase the chances of a discovery. More important than a very dark and clear sky, I reasoned, was the need of a site with more frequent clear nights. The Canadian sites from which I had done most of my observing in the 1960s and 1970s seldom offered more than two clear nights in a row at any time of the year except for a week or two in May and again in September.

Certain times of the month that are crucial for comet hunting: right after a full moon and around the new moon. If I wanted a better statistical chance of discovering comets, I needed a site with a greater likelihood of recurrent clear nights during these periods. Therefore in 1979 I decided to move to Tucson, Arizona, where some 300 nights a year are clear enough to stalk comets. I chose the community of Corona de Tucson, some 20 miles east of town, where the sky is pretty dark. I built a small observatory out of a 9 × 10 foot garden shed, designing the structure so that the roof slid off to reveal the open sky. Within a year I had a 16-inch diameter reflecting telescope, and then I resumed comet hunting in a big way.

A comet in spring 1983 attracted my attention because I should have found it myself. It was discovered by *IRAS*, an amateur astronomer G. Araki from Japan, and George Alcock, a nova and comet hunter from Peterborough, England, some 60 miles from London. Although the IRAS satellite detected the comet first, the

Comet Bennett, March 1970. (Photograph by Jack Newton, using a 450-mm telephoto lens; 2-minute exposure using Tri-X film.)

people operating the satellite did not immediately know what they had found. Meanwhile amateur astronomer Alcock was in his pajamas, searching for exploding stars when he saw the large comet through a pair of binoculars and a closed window.

A month later I actually did find a comet. After a brief flurry of excitement, I checked the pages of *Sky and Telescope* magazine and learned that my new comet was actually a well-known old one, Periodic Comet Tempel 2. Then at the end of November, I came very close. This time it was Comet Hartley–IRAS, 1983v. Discovered by the IRAS satellite from space and by Malcolm Hartley, using a large wide-angle Schmidt camera in Australia, this comet was supposed to be a faint fifteenth magnitude, far too faint for my telescope. However I had not heard of the report when I picked it up the comet at the end of a 2-hour evening comet sweep. So I had independently discovered a new comet less than a week after it was first reported. At least I was finding things! I began to think that at last my long wait might be over. Although I was pushing 900 hours of comet hunting, I was nowhere near the record set by California amateur Don Machholz, who spent 1700 hours with his eye at the eyepiece before he found his first comet and another 1700 hours before his second!

My first real discovery took place on November 13, 1984, as I described in the preface to this book, but it was quite a while before my second find. Meantime Halley's comet—the most famous of all—was due back, creating lots of excitement both within the astronomical community and among the general public. Starting in 1985 I joined Steve Larson as part of the Near-Nucleus Studies Network of the International Halley Watch. During the 3 months before perihelion passage in February 1986, we observed the comet every night with either a 20-inch or a 61-inch telescope and an electronic charge-coupled device (CCD) detector system. Our goal was to study the area around the comet's nucleus. We recorded the complex structure of dust jets that erupted frequently, changing the comet's appearance from one night to the next. On January 6, 1986 we saw a dust jet going straight in the direction of the tail, an almost identical repeat of what the astronomer George Ritchey had photographed from Mount Wilson during Halley's previous visit in 1910.

With Halley's comet far in the southern sky in spring 1986, we discontinued the nightly CCD work from Tucson, but in April, I did get to go to Peru to lead a Halley's comet expedition. On the first night, I was scouting our observing area for the best place to watch—just me and Minerva, my 6-inch telescope. It was a remote site on the grounds of an Inca ruin. Surrounded by tall stone structures and a dark sky, the site seemed delightful. Not only was this my first night in a strange country, it was my first night ever in a new hemisphere. I set up the telescope and started looking at Halley's comet, which was just rising.

As I was concentrating on the comet, I didn't hear some faint distant footsteps until they were loud enough so that I could tell somebody was approaching. The steps got louder; there was the clinking of glass. There was conversation in a language I had never heard before, but it was clear that my uninvited visitors were somewhat soused. Not knowing what to do, I looked at the distant comet and then at Minerva. Hmmm, I thought, telescope as a weapons system. If they got threatening, maybe I could hit them over their heads with my telescope.

As the men got closer—in the starlight, they seemed to be awfully big hombres—my fears were confirmed. There were two, they were drunk, and they were staring straight at me. For a minute, the three of us stood there. I looked at them, they looked at me, at my telescope, then at the sky. Then they pointed. "Hal—leee?" I offered them a turn at the eyepiece. We ended up having a good time comet watching together, although I suspected everything looked fuzzy to them that night!

COMET HUNTING BY READING

That same year I interviewed Clyde Tombaugh, discoverer of Pluto, for a biography I would eventually write.[3] I asked Tombaugh what other objects he had found, and as he listed the clusters of galaxies, star clusters, variable stars, and asteroids, he let on that he had also found a comet. "But I've never heard of any Comet Tombaugh," I protested.

"I found it on plates taken many months earlier," Tombaugh explained. Knowing the comet was long gone, he figured that it would be impossible to follow it and therefore there would be no sense reporting it. "Would it be possible to see the image of that comet now?" I asked hopefully. Tombaugh said that the detailed notes he had taken on each plate were still preserved at Lowell Observatory in Flagstaff, Arizona, so all I had to do was locate the correct plates. Then finding the comet on them would be easy. "Great!" I said. "When did you take the plates?" "Sometime between 1929 and 1945," he replied impishly.

February 9, 1986—the perihelion day of Halley's comet—found me in the subterranean plate vault at Lowell. The room was furnished in a spartan way with an old wooden chair and table. All the old plate envelopes on which Clyde had recorded his meticulous notes had been replaced with modern archival ones, but the old ones were still filed away. I took out several boxes of them and set to work. On each big brown envelope, Tombaugh had listed the numbers of variable stars, asteroids, comets, and other unusual objects. On envelope after envelope, I saw a procession of "no comets —no comets" recorded in his scrawling hand.

After about an hour, Arthur A. Hoag, director of Lowell Observatory, came by to visit. He looked at me and the large boxes of envelopes spread around. "Now that's a new one!" he smiled. "Comet hunting by reading!"

When I had progressed to the middle of February 1930, I picked out the envelope for January 23, which listed the usual notes, then at the bottom: "4. No comets. 5. Planet X (Pluto) at last found!!!"[4] Interrupting my own search, I carefully removed, examined, and replaced the historic first discovery plate of Pluto. Then I went back to work. It was already late in the afternoon by the time I reached the end of January 1931; still "no comets—no comets— no comets," then "Comet?"

Was that it? A quick check of the plate showed the telltale trailed image of a comet sitting on the plate like a faint smudge with a tail on it. But this one had never been reported, never before announced. The soft image, recorded black on white on the plate, was a comet that passed by without revealing its secret till it was gone, leaving only a few photons of light frozen on the photographic emulsion. Even now only two people knew it existed, Clyde and I, and I was the only one who knew where it was!

Now with the help of my friend Brian Skiff, an observer on the Lowell staff, we set out to measure the comet. Then Skiff found that Lowell Observatory had actually reported the object back in 1931 as an asteroid, not a comet, and it had been given the asteroid designation 1931 AN. When we reported the object—correctly as a comet and with accurate positions—to Marsden at CBAT, he said that he would not be able to announce the comet unless an accurate orbit could be determined. And for that I needed more observations. That summer I went to the gargantuan plate collection at Harvard College Observatory, but Comet Tombaugh was too faint to appear on any of the plates in its collection. Over the next 2 years, whenever I had the chance and money to travel, I visited plate collections as far away as Heidelberg in Germany, Mount Wilson in California, and other collections around the world. Heidelberg had a plate so close that I was sure the comet would be there; sadly it was just off the edge. Mount Wilson also had a plate that was close. Of course had the comet appeared on either of these plates, someone might have discovered it at the time.

The moral of this tale? Comet hunting by reading would have worked if I had found more observations of this comet. Although Marsden is pretty sure that Comet Tombaugh was never seen before and it is possibly periodic, without more observations he can't be positive. A year later Skiff found that Tombaugh had discovered a second comet, so now we had two unannounced comets. Will either comet eventually be announced as Comet Tombaugh?

BIGGER AND BETTER COMETS

The next year I snared two new comets, but both were close to the horizon at discovery and found under difficult conditions. The first appeared only 5 days into 1987. I had not found a comet in a while, but I was also on a tight deadline for my book, a guide for observing variable stars. So on New Year's day 1987, I made two resolutions: one to find a comet and the other to finish the book. On the morning of January 5, after working hard all night, I finally typed the last sentence of my book and started to print it. The sky had been very cloudy all night in advance of a storm. Each time I checked, it got cloudier and cloudier, so when I started printing the manuscript, I was not expecting to see that a clear spot was beginning to open up in the eastern sky. As fast as I could, I turned my house from a writing office into an observatory. I turned off white lights and turned on red ones so that my eyes would adapt to the darkness outside, and I walked into the back yard. There I had a garden shed that I had made into an observatory some 5 years earlier. These metal sheds are set up so that the roof supports the walls, but for my purpose, the roof had to slide off on rollers. It was quite a challenge making wooden supports to turn my garden shed into a place for comet hunting.

Out in the garden shed observatory that morning, I grabbed a handle and gave a shove. Slowly the big roof started moving, and a minute later the observatory was open to the sky and I began a slow eastern sweep with Miranda, my 16-inch diameter reflecting telescope. But the search was interrupted by both the threatening clouds and my need to check the printing of my book from time to time. After my second printer check, I asked myself what would stop this hunt, the clouds or the dawn. It was neither. I returned to the telescope, moved one field to the east, and there was a comet.

I had barely 45 seconds to observe and record this object before clouds and dawn rushed in. The position I had estimated was therefore very poor, and besides there was a faint object marked on the atlas that I keep for handy reference, so I assumed that it must be just a galaxy of some sort. By now the sky was quite bright, and as I looked outside into oncoming day, I was glad I had remembered to close the sliding roof observatory, because now it was raining heavily. I remember how exhausted I was, having been up all night writing and then observing. As I looked toward the now-closed observatory and the rain, I asked myself if I really had discovered a comet in the last hour.

By the following morning, the sky had still not cleared, but I did check the best sky map of all, the Palomar Observatory Sky Survey and found that the faint object on the other atlas was not the object I had seen. By the morning of January 7, the sky had cleared at last, and I quickly found the object some distance from its earlier position. Marsden then announced it as Comet Levy, 1987a. Within the first week of the new year, I had finished my book and found a comet—I had achieved both of my New Year's resolutions.

In spring 1990 the world was waiting for a bright comet. Found by Rodney Austin from New Zealand at the end of 1989, Comet Austin $1989c_1$ promised to become pretty bright as it swung around from the south. At that time, I was observing with Steve Larson from the Lunar and Planetary Laboratory in Tucson, and we had planned to spend a week at the large 61-inch telescope in the Catalina Mountains north of town observing the comet in different filters and obtaining spectroscopic images of its light. But by the fifth night of our run, it was clear that Comet Austin, though a pleasant object, was not nearly so bright or so active as we had hoped. David, Steve said early in the morning of May 19, "Comet Austin is a one-person job. Tomorrow night you should stay home and find us a bright comet!"

With morning sky washed out by a last-quarter moon, I decided to accommodate Steve by hunting earlier than usual. I opened the roof and began hunting around 2:30 A.M.; I had intended to continue until moonrise, but patchy clouds were interfering even with that plan. Then the sky improved as the moon rose an hour later, so I decided to keep on searching for a while. Swinging the telescope north to keep as far from the moon as possible, I hunted through the constellation of Andromeda, past the bright star Alpheratz, and swept on into Pegasus.

Not far from that bright star, my telescope stopped, and my brain snapped out of its lazy state as a soft fuzzy patch of light entered the field of view. Since I knew this part of the sky very well, I quickly thought that this patch must be a comet, but it did not move over the next hour. A check of the Palomar Sky Survey showed nothing in that position. Twenty-four hours later, the comet had crept along barely an eighth of a degree, by far the slowest motion I had ever seen for a comet at discovery. Announced as Comet Levy 1990c, this new visitor was far from the Earth and heading toward us. Plodding toward Alpheratz, a star in the Great Square of Pegasus, the comet seemed to be preparing for its onslaught of the Earth's sky.

By early July Comet Levy was picking up speed and brightness as it approached both sun and Earth. In late July while visiting my family in Montreal, I went south of town to show the comet to a reporter for the Montreal *Gazette*. The sky had been cloudy for a few days, so I was not sure what the comet would look like. But when we arrived at the site, I looked up and saw the comet visible to the naked eye, and through Minerva my 6-inch diameter portable telescope, it was a marvelous sight. By the middle of August, the comet cut a striking path right near the summer Milky Way. That was the year I was the main speaker at the annual Stellafane Telescope convention in Vermont. The talk was in a large natural amphitheater, and my audience consisted of 1500 people and one comet rising in the east.

Seven months later the comet was still a striking sight. In February 1991 it crossed the plane of the Earth's orbit and for 2 or 3 weeks sported a beautiful antitail. Composed of sunlit particles around the comet, the antitail points toward the sun instead of away from it and becomes visible when the Earth crosses the plane of the comet's orbit. In May 1991 I observed my comet on the first anniversary of its discovery. Its show was over; it had faded a great deal and was hardly visible at all.

Comet Levy 1990c. (Photograph by David Levy using the 18-inch Schmidt camera at Palomar, August 24, 1990; 8-minute exposure.)

A NEW PERIODIC COMET

A week later, with dawn less than an hour away on the morning of June 10, I awoke to search the morning sky. But as usual for that time of year, it was still cloudy. Not wanting to give up

Comet Levy (1990c). (Photograph by David Levy, January 1991, using the Palomar 18-inch Schmidt camera.) Since the Earth was passing near the plane of the comet's orbit, dust near the comet is visible as an "antitail" that points downward in this photograph.

just yet, I opened the sliding roof on my garden shed and went back inside to watch a "M*A*S*H" rerun. After the episode had finished, I stepped outside. The sky was looking a bit more promising, but it was still far too cloudy. I checked the sky a final time, for by now the first signs of dawn were appearing. There were still a lot of clouds, but growing patches of clear sky made me decide to begin comet hunting low in the eastern sky. I decided to search in the constellation of Aries, which was just coming up over the trees.

The telescope's field of view was about 1 degree across, or two moon diameters, and full of faint stars. I found nothing fuzzy or unusual in that field, so after a few seconds, I moved the telescope eastward and checked the adjacent field of view. After a minute of slow sweeping in this way, I moved the telescope again. Now a bright fuzzy patch of light entered the field of view, and for a second, a now-familiar "red alert" went off in my brain. But this fuzzy patch didn't fool me for long: By its elongated shape and its position, I knew that this was a distant galaxy called Messier 74. This galaxy is a highwayman, guilty of stopping comet hunters dead in their tracks. More than 200 years ago, it duped Messier, who sketched its position and then checked back later to see if it had moved, as all comets do. When the object remained frozen in the sky, he added it to his catalog as he went on in search of slowly moving cometary prey.

I've encountered Messier 74 so many times that I know it as an old friend. But the sky was brightening, so there was no dawdling over M74 now. I nudged the telescope down a field and another and another, slowly moving toward the horizon. When it got low enough to see distant treetops, I turned the telescope southward and then started sweeping back upward along an adjacent track. Another minute passed by, and then the mental alert went off again: There was another fuzzy spot.

For an instant I thought it was Messier 74 again, since the galaxy was close by. But wait a minute: This object was quite a bit brighter. With mounting tension I put in a higher power eyepiece and looked more closely. While M74 had relatively sharp edges all around, this thing had a bright center, then faded off so slowly that I could hardly tell where it ended and the sky began. Sharp edges are characteristic of a galaxy filled with stars; the gradual fading is a comet's typical signature. This, I decided, was a comet.

It was my seventh discovery from my back yard. I marked the comet's position not far from M74 on the star atlas. Then putting the telescope on two stars whose brightness I knew and changing the telescope's focus so that they appeared fuzzy, I estimated the comet's magnitude as a bright eighth magnitude. Although that is about five times fainter than the faintest naked-eye star, through my 16-inch telescope the comet was quite distinct. A closer look, and I noted a short tail pointing away from the sun.

Finally with the sky now so bright that the comet was virtually invisible, I plotted its position again. It had moved a tiny minute of an arc—only one-thirtieth the diameter of the full moon—in the 20 minutes since I had first seen it. It was time to report. Armed with all this information about an object whose existence might still be a secret to the world, I went inside to compose a message to Marsden. An hour or two later, a one-page announcement circular appeared at observatories and universities around the world.

But what was the orbit? Where was the comet coming from? Was it heading in, like 1990c had been, or had it already rounded the sun? Was it visiting the inner part of the solar system for the first and only time, or was it a new periodic comet that would return on a regular schedule? To answer these questions, astronomers take lots of photographs of the comet, either on film or electronically, and then measure precisely where the comet is each night as it crawls across the sky. With that information, Marsden could calculate the comet's obit. Within a few days, another CBAT circular appeared with a preliminary orbit for the comet, but it was a full month before enough accurate positions from telescopes all over the world told a better story. Curving around the sun in just a bit over 50 years, Comet 1991q is now called Periodic Comet Levy, a newly welcomed permanent member of the sun's inner circle. What's more, the orbit is similar to that of a bright comet that appeared almost 500 years ago when Columbus was discovering America. Determining whether the new comet is actually a return of the comet of 1499 will have to wait until the comet rounds the sun and appears again in 2042. Until then we simply will not know the orbit well enough.

A NEW LOOK AT COMETS

For any comet hunter, each find has its own fond memory. Like my comets of 1990 and 1991, some finds come easily, with the suspense-filled moments from discovery to certainty taking less than an hour. With the comet I found in 1988, the uncertainty lasted a lot longer, but it was worth the wait, for it was the comet find that year that led me to Eugene and Carolyn Shoemaker and a whole new way of looking at comets.

Early in 1988, my friend Steve Edberg encouraged me to accompany him on an expedition to see the total eclipse of the sun on March 17. He wanted me to join a special boat charter to a site where the eclipse would be total. The cruise sounded like fun, but since it would happen right in the middle of the best time for sweeping comets, I said I would pass. "Why can't total eclipses of the sun," I asked, "take place when the moon is in waxing gibbous phase? You know, a few days after first quarter? Then it wouldn't interfere with dark sky observing."

"You're probably not going to find a comet this one time!" Steve protested. But there was also a meeting on asteroids going on in Tucson that I didn't want to miss. No comet hunter should be indifferent in these comet cousins. And besides Eugene and Carolyn Shoemaker, who together had discovered 17 comets, would be at the Tucson conference, and I very much wanted to meet them.

It was a very successful meeting, and the Shoemakers and I had a chance to become familiar with each other's methods for patrolling the sky. Even though their methods were photographic and their main targets not comets but asteroids, we found that our work had quite a bit in common. I left that meeting invigorated.

New moon and the eclipse approached less than 2 weeks later, and on the morning of March 19, about a day-and-a-half after the eclipse, I was searching through the constellation of Pegasus when I saw something that looked like a galaxy. I was about to throw it back into the sky like an unwanted fish—it was symmetrical on both sides, like a long spiral galaxy—but it seemed a tiny bit suspicious, so I sketched its position and went on. The next morning was also clear, so I resumed searching even though my asthma was making me a bit wheezy. But when I checked the position of the previous night's galaxy, it wasn't there! Had I marked the position incorrectly? But half a degree to the north, there was a very different-looking galaxy. While yesterday's galaxy was evenly diffused over its whole surface, this one had a bright, almost starlike central core.

Quite confused by what I was seeing, I first wondered how I could have been so far off with the first object. I was skeptical about the whole thing, and not trusting my instincts, I decided to give it another day. The third morning was also clear, and when the object rose, I was ready for it. I pointed the telescope toward the first night's position; no galaxy there. Then I moved to the second night's position. The central core was still there, but the rest of the galaxy had disappeared.

Now I had the answer. This must be a comet, unusually elongated, that passed right in front of a star on the second morning. To confirm my hunch, all I had to do was move the telescope one-half degree to the north. I was about to push the telescope when the thought hit me: This was an important moment; my next motion of the telescope should, if I were right, yield a brand new comet. I took a deep breath, grabbed on to the telescope, and moved it one-half degree to the north. There in the middle of a circlet of stars lay my new catch. "Aha," I said, "I have you now."

By now I should have known what to do. I should have been a picture of serenity as I sketched the comet's position, opened an atlas of the sky, and plotted it, then estimated its brightness. But it didn't happen quite that way. Comet hunting is normally one of the most relaxing of pasttimes, but when a discovery occurs the atmosphere changes like still air cut by lightning. Hardly serene, my mind was a mass of quivering impulses as I paced around the telescope without the slightest idea of what to do next.

Other comet finders tell me that they also get carried away by the excitement of the moment. But this morning I couldn't afford that luxury. There were a dozen or so active comet hunters throughout the world who at that very moment were in search of the Holy Grail that I had just found. Come on, David, get with it, I told myself; now is not the time for a holiday.

Quickly logging into Marsden's CBAT computer service, I reported the three nights of positions and the comet's faint eleventh magnitude. Feeling triumphant but exhausted, I went to sleep. A few hours later I awoke, logged into the service once more and saw that Comet Levy 1988e had already been announced. But the story of this faint little visitor was far from over.

On March 22, 1988 Eugene and Carolyn Shoemaker were getting ready for a night of searching for asteroids and comets with the 18-inch Schmidt. It looked like the sky would stay clear, so with exposures lasting 8 minutes each, they would take as many as 40 exposures that night, each one on a 6-inch diameter piece of black-and-white film.

Like all photographic hunting expeditions, the Shoemaker search game is played quite differently from that of the visual searchers. Instead of hunting the sky with eyepiece and telescope, the husband-and-wife team captures comets and asteroids on the emulsion of photographic film. With the first film of the night safely loaded into the telescope, they move the telescope to a new location, then center the telescope on a star. After a countdown, two metal shutters squeak open at the top end of the telescope.

When the Shoemakers heard of Comet Levy 1988e, they decided to try to photograph it to help determine its orbit, even though the comet would not be high in the sky before morning twilight began. On that night, they finished their regular program late, so they quickly moved the telescope toward the east and took a short exposure of the comet field. To their surprise, despite the brightening dawn sky they got a good image and submitted the first accurate positions of my comet.

At their next observing run, the Shoemakers, now accompanied by their colleague Henry Holt, again put my short-tailed comet on their observing schedule. This list is set up graphically, with a sheet of paper on which dozens of nickel-sized circles outline the observing fields around the sky. But since the comet was so far north of where they usually observe, the extra circle would have to be placed way off the top of the "nickel diagram." Instead, Shoemaker simply inserted an extra circle at the top left-hand side of the page.

On the morning of May 13, the Shoemakers finally turned their telescope to the position indicated by the extra circle. But this was 5 days into an exhausting week of observing. They turned the telescope to the position indicated by the circle, and took a set of exposures. Carolyn did not get to scan those films until the start of the seventh and final night of the observing run. Gene and Henry were about to leave to set up the telescope upstairs.

Carolyn quickly found what she was looking for. "Gene," she called, "would you like to see David's comet?" Gene came downstairs and looked through the stereomicroscope. "Why is the comet so far off the center of the field?" he asked, and then added that it was brighter than they expected. Gene then went back upstairs to continue setting up for the night's work while Carolyn wondered why the comet was not at the center of the field as she had planned. After a few minutes, she called upstairs again: "Gene, I cannot identify any stars on this film. I don't know what field this is!" Shoemaker was already guiding on a star at the telescope. "Henry," he said and quickly handed over the guide paddle, "Take over." After he rushed downstairs, it took him a few minutes to identify the mistake: They had taken the field represented by the position of the circle on the nickel diagram, not the position where Comet Levy was—a difference of 17 degrees. Satisfied that he had solved the riddle, Gene was about to go upstairs again when Carolyn asked the question, "But if this comet isn't David's, then whose is it?"

A quick check showed that it would be theirs, since there was no known comet at all in that position. Before the night was over, they had reported their quite accidental discovery of Comet Shoemaker–Holt, 1988g. I heard about this discovery after I had returned from a star-observing party in Texas. The comet seemed bright enough that I might be able to detect it in the morning sky using my own 16-inch telescope. Sure enough, I was able to see a faint fuzzy patch. As the sky brightened at the approach of day, I closed up the telescope and went back to sleep.

A few hours later the telephone rang. It was my friend Charles Morris, an experienced comet observer who lives in California. "David," he wanted to know, "did you see Comet Shoemaker–Holt last night?" "Yes. Can't this wait? I am still asleep." "Have you seen the orbit?" he interrupted breathlessly. "Conrad Bardwell noticed this yesterday. Comet Shoemaker–Holt's orbit is almost exactly the same as Comet Levy's except that it comes to perihelion 3 months later!"

Then associate director of the Minor Planet Center, Conrad Bardwell, was working with Marsden at that time to calculate orbits for asteroids and comets.[5] As soon as enough positions were available for the new Shoemaker–Holt comet and a rough orbit calculated, Bardwell noticed how similar its orbit was to that of Comet Levy. Inclined some 63 degrees to the plane of the ecliptic, the two orbits were almost identical in every respect except that Shoemaker–Holt had arrived at perihelion on February 13, 1988, about 3 months after Comet Levy's closest approach to the sun on the nineteenth of the previous November. The orbits were too similar to be a coincidence. More likely this was the first instance, aside from the various sun-grazing comets, of a pair of related long-period comets being discovered independently.

When more positions were available for both comets, the calculated orbits became almost identical. At the end of 1988, Marsden suggested a scenario: Some 12,000 years ago, a single comet broke in two as it rounded the sun. The two parts did not separate right away but stayed together as a double comet until millennia later and far from the sun, they began to drift apart.[6] It is lucky that I went comet hunting instead of eclipse chasing that eventful night of March 19, 1988. The first comet was faint and close to the dawn sky. I observed it for three nights before reporting it, and there was no independent discovery by another observer—in fact there is no record of any independent find. The other comet was far from the Shoemakers' usual search area, and they would have missed it had they not been trying to record the first comet.

It is often from accidents like these (through serendipity) that interesting scientific discoveries are made. Moreover the confluence of visual and photographic comet searching has enhanced our abilities to make a find. Comet Levy 1988e and Comet Shoemaker–Holt 1988g are a remarkable pair of comets, but for me, the co-incidence was even more important because it led me to the Shoe-makers, a whole new way of finding comets, and a whole new way of looking at comets. Instead of just being beautiful objects in the sky, I now saw comets as forces that could plow into the Earth and even change the course of life here.

❨ 9 ❩

An Asteroid Hits the Earth

Imagine living in Flagstaff, Arizona, the day that Meteor Crater was made. A blinding flash of light in the eastern sky startles you; you look toward it. There is a billowing cloud of debris rising into the stratosphere and mushrooming out. But you don't hear anything. Like a scene in a silent movie, the events 45 miles away seem somewhat detached, as in a dream. Three minutes and a few seconds later, the illusion is shattered with a glass-breaking thunderclap. Then a low rumble goes on and on, and the ground trembles.

But all this happened some 50,000 years ago, possibly long before the first human beings explored the new world, longer still before they came across the flat-topped formation we now call Coon Butte and its huge hole in the ground—200 meters deep, about 1.2 kilometers across, and made in 5 seconds by an asteroid only 50 meters wide that smashed against the Earth.

Nowadays the crater's climate has all the harshness of its mile-high desert location, hot by day and cold by night. But on the day the asteroid hit, the Earth was in the midst of an ice age, and Northern Arizona's climate was wetter and cooler than now. The object careened into a thick pine forest, before an audience of mammoths, ground sloths, bison, large lions, and sabertoothed tigers. For thousands of years, this huge scar just sat there, part of the landscape, and slowly eroded, until 1886 when shepherds watching their flock on the plain surrounding the crater discovered some pieces of iron. In a pretty ingenious bit of reasoning, the shepherds figured that a single explosion had somehow expelled the iron as it formed the crater.

Grove Carl Gilbert, one of the top geologists in the country at the time, was the first to study the crater scientifically. But his hunch that the crater was the result of an iron-rich asteroid impact didn't pan out in the tests he conducted. He hoped that a huge mass of iron beneath the crater floor would cause an anomaly in his compass reading. It didn't, so he assumed that if the crater were the result of an impact, then the object must be too far below the surface to cause a magnetic blip. Gilbert described the failure of his test in his retiring presidential address at the Geological Society of Washington from a philosophic point of view.[1] The paper discussed how a hypothesis is born, figured through, and then either accepted or discarded. Meteor Crater served the purpose of an example. "The mental process by which hypotheses are suggested," he wrote, "is obscure. Ordinarily they flash into consciousness without premonition." Gilbert proposed that hypotheses come out of analogy: To explain an unusual event or a peculiar feature, the scientist considers what aspects are already known, then builds the hypothesis on that.

Gilbert's paper addressed the two likeliest ways that a crater is formed: by the explosive eruption of a volcano or by the impact of an object from space. For craters here on Earth as well as on the moon and other planets and moons, scientists are forever debating which giant pockmark was caused by the one and which by the other. In fact Gilbert gave the first clear exposition on lunar craters being formed from impacts with other objects in space. "What would result," Gilbert then asked, "if another small star should now be added to the earth, and one of the consequences which had occurred to me was the formation of a crater?" Gilbert thought that the main body would be composed of iron and if it lay beneath the crater, it would cause "a local deflection of the magnetic needle."[2] He noticed that the crater 45 miles east of

Flagstaff was not circular but squarish. Perhaps the star "struck the earth and bounded off, finally coming to rest at some point further east?" But he found that the crater's shape was inconsistent with an asteroid hitting and then ricocheting.

Gilbert concluded that the crater was formed by a steam explosion, like what happened to Mt. Vesuvius in A.D. 79 or to Krakatoa in 1883. After all the crater is less than 10 miles from a known volcanic crater. But—and Gilbert noticed this too—most volcanic craters are on top of their mountains; this one bore only a casual relationship to the butte it indented. And what about the nearby meteorites? Could the meteorite fall have "touched the volcanic button," setting off a volcanic explosion?

THERE'S PLATINUM UNDER THAT THAR HILL

In 1903 Daniel Moreau Barringer, a successful mining engineer and lawyer, heard about the crater and the platinum-laden meteorites surrounding it. He believed that the main body of the meteorite, still hidden under the crater, must be loaded with the stuff. His agents went out, staked the claims, and then sight unseen, he bought the land. He put half a million dollars into exploring for what he thought would be a big mass of meteoric iron under the crater. Much of the rest of his life was spent in a search for this main mass of iron and in a fruitless attempt to mine it. The government later established a post office in the area and named it Meteor, not for the crater but for the meteorites surrounding it.

Barringer's dream was never fulfilled. He died in 1927, one year before Eugene Shoemaker, the man who would finally figure out Meteor Crater, was born.

METEOR CRATER ON THE ROAD TO THE MOON

When Muriel Shoemaker gave her 7-year-old son a set of marbles in 1935, she thought she was simply passing on some unusual family keepsakes. But these marbles, some of which were real agate, fascinated him. A year later during a trip with his father to South Dakota's Black Hills, Shoemaker was so impressed with the rose quartz and other minerals in the area that he began to collect them. His dad knew just enough about them to whet Shoemaker's appetite for more. That year their family separated briefly when his mother took a teaching position at the School of Practice at Buffalo State Teachers College in western New York, while his father worked at the Civilian Conservation Corps in Wyoming.

Fresh out of fourth grade, Gene enrolled in an unusual program in science education for children. It included field trips in geology and evening courses in several sciences—quite an opportunity for a child. Combined with the fossil-rich geological formations on Eighteen-Mile Creek, south of Buffalo, the program's 5-year duration left Gene convinced that he wanted to be a geologist. With his family planning to reunite in Los Angeles where his father had found a job as a grip on a Hollywood sound stage, he graduated from ninth grade and moved West in 1942.

Gene brought his growing love of minerals to Fairfax High School in Los Angeles, where his activities on the gymnastics team gave him his later goatlike agility on field trips. Although he played violin in his high school orchestra, he was pretty single-minded about the direction his career would take even then. He decided to spend his first Los Angeles summer working as an apprentice lapidary. He enjoyed cutting and polishing stones so much that his parents worried that he might never go to college. But their belief in the value of a higher education—preferably at Caltech—was not lost on him.

Gene Shoemaker started at Caltech in fall 1944. To train navy engineers, the school had adopted an intense wartime routine in which the year was divided into three semesters instead of the usual two plus summer vacation. "I squirted out of Caltech in 2⅔ years," is how Shoemaker describes his undergraduate career and graduation at the young age of 19. He stayed on at Caltech to earn his master's degree in 1948, this time after having completed a thesis about the petrology of a small area of Precambrian meta-morphic rocks in northern New Mexico.

That year Gene Shoemaker joined the US Geological Survey to complete an important project: searching for deposits of uranium in western Colorado. At the dawn of the nuclear age, the United States had very limited knowledge of where uranium deposits were. Headed by geologist Richard Fisher, this project was vital both for the production of nuclear weapons and for the fledgling nuclear power industry. Those were the days when we were told that nuclear power would be too cheap to measure—but first we needed a supply of uranium.

STARTING A DREAM

Out in the western Colorado boondocks that year, Shoemaker was sure that human beings would land on the moon in his lifetime. "It all came to me in a flash one morning while driving to breakfast," he recalls. Only 20 years old, he was living in a mill camp in West Vancoram, Colorado, preparing for a diamond-drilling project. He had already made up his mind that the Earth would be his laboratory, but on that morning, the Earth was suddenly no longer enough.

"I had my meals 5 miles away in Naturita, down at the head-quarters of the Vanadium Corporation of America," he reminisces.

> I was driving along the road along this beautiful river. Aha! That is what I will aim to do: to be one of the first people on the Moon. Why will we go to the moon? To explore it, of course! And who is the best person to do that? A geologist, of course! I took the first fork that went to the moon that morning.

If that daybreak drive saw the start of Shoemaker's road to the moon, joining the US Geological Survey that summer was his first turn. "I didn't tell anybody about it at the time—it was a secret goal and it might have been considered silly," he said. Since craters are the most dominant feature on the moon—even the smallest telescope will show at least 300 of them—early on that road, his thoughts turned to craters. How did they form? Were they volcanic? Were they the result of comet or asteroid crashes? "I deliberately studied those aspects of geology that I thought might be turns on the road to the moon," he says. "I studied violently explosive volcanoes—a turn on the road to the moon." After 2 years in the field, he was accepted at Princeton University for his Ph.D.

THE LANGUAGE OF THE EARTH

At this time in his career, Shoemaker's goal was to go into the study of igneous and metamorphic petrology—the study of rocks formed by volcanism or changed by heat and pressure—and he looked forward to working under Arthur Buddington, one of the nation's top petrologists. Since his early weeks at Princeton were slow, Shoemaker decided to use them to dispose of his language requirements: French and German. Designed to give Ph.D. candidates the ability to read journal articles in the world's major scientific languages, the requirement was handled not by language departments but the geology department itself. French was taught by the department's invertebrate paleontologist, Benjamin Howell. Shoemaker got a copy of Eduard Seuss's *Le Face de Terre*, a famous turn-of-the-century summary of the geology of the Earth printed in French, German, and English. About a week later, he returned to ask Howell a routine question. Howell rose from his desk and fetched a book written in French. "Translate this!" Howell said. With some surprise Shoemaker began translating the paper orally. Howell provided a few missing words, and in less than a half-hour, Shoemaker passed his French requirement. "Not bad!" Shoemaker thought. "Now I'll try German." Shoemaker then went to the German version of the book, *Das Antlitz der Erde*, but finishing German took longer.

Game plan is one of Shoemaker's favorite expressions, and for good reason. Since early in his life, he has always had a clear idea of where he wanted to go. For his doctorate, the Shoemaker game plan was to finish up completely in 2 years. Moreover in the 2 years after leaving Caltech, he had completed most of the field work for what he expected to be his dissertation, the geology of the beautiful Fisher Valley–Sinbad Valley salt structure that spans the Colorado–Utah border. So in the summer of 1951, Shoemaker returned to Grand Junction, Colorado, for two reasons: to prepare some geologic quadrangle maps of his thesis area for publication and more importantly, he hoped to marry Carolyn Spellmann, a young woman he had met a few years earlier.

Gene first met Carolyn when he was best man at her brother Richard's wedding. Gene and Richard had roomed together at the Caltech. Although Carolyn had a boyfriend, Richard wanted to introduce her to Gene. "If you will give us some time," he bribed his sister, "I'll give you this wonderful tablecloth." Carolyn agreed. "I was easily bribed," she admitted.

Born in 1929 in Gallup, New Mexico, Carolyn grew up in Chico, California, at the northern end of the Sacramento Valley. Her father was a poultry farmer and her mother a schoolteacher. Carolyn got her A.B. and M.A. degrees from Chico State College, now the University of California at Chico.

FROM URANIUM TO IMPACT CRATERS

Gene Shoemaker fully expected to return to Princeton in the fall of 1951, write his dissertation, and finish his Ph.D. on schedule. But before that summer was over, the geological survey's directors turned Shoemaker's plan on its head by calling on him to return to studying uranium deposits and the regional pattern of their chemistry. The prospect excited him. Thinking that there may be clues in this pattern that would lead to important discoveries of ore, he set out on an adventure. This one small change in his game plan is what led him to have an impact on craters and space.

Gene Shoemaker's road to space began in a field of ancient volcanoes in northern Arizona's Navajo reservation, called the Hopi

Buttes. In the course of his regional investigation, Shoemaker had discovered uranium deposits in the eroded volcanic vents of those ancient volcanoes, and now he had the chance to study them. "I already had that in mind as a stepping stone to the moon," Shoemaker remembers of his time spent there mapping the eerie-looking structure of the ancient volcanic vents. "I wanted to understand how violent volcanoes work. They actually make craters that in form are quite similar to some lunar craters." At the time he suspected that some lunar craters might have been created by the same process.

In fall 1952, returning with Carolyn and a colleague from a field trip to the Grand Canyon, Shoemaker paid his first call at the crater on Coon Butte near Winslow, Arizona. The three exulted in the prospect of seeing Meteor Crater as they rumbled along Route 66 in an old government-issue jeep station wagon. When they arrived, they found that it cost money to get in. "We had just enough money to get some fried rice for dinner," Carolyn remembered. "We were down to nickels and dimes." With a peek over the rim at sunset, they got a sneak preview of their future occupation.

THE FINAL SEARCH FOR THE CRATER'S ORIGIN

Meteor Crater was obviously the result of some sudden convulsion, but by 1954, Shoemaker still wasn't sure what had caused that convulsion. Could it be a salt dome that had collapsed or even a volcanic explosion? Or could it be the result of a hit from an asteroid from space? In 1954, a paper by geologist Dorsey Hager turned Gene's mind back to Meteor Crater and started his final sprint to figuring out its cause. The paper described how petroleum drillers had discovered structures called evaporites a few tens of miles east of the crater. Since the crater was nearby, Hager suggested that perhaps it was the result of a collapsed salt dome, but Shoemaker was skeptical. The crater, he thought, is in a pretty busy geological area; in addition to the evaporites, it is also near a region filled with ancient volcanoes. It was just as possible that the crater could be a Maar-type volcano similar to volcanoes he was studying in the Hopi Buttes.

Shoemaker and Hager started writing to each other, and Hager sent a sample of pumiceous silica glass by mail. Years earlier when digging shafts beneath the crater floor in his search for the huge mass of iron, Barringer had found this glassy material. To test whether this glass were volcanic, as Hager suggested, or melted sandstone as Barringer had thought, Shoemaker sent the sample for spectrographic analysis.

The result: Both the salt dome and volcano ideas were blown out of the water. The glass had the same composition as the Coconino sandstone he had sampled in the Grand Canyon, and the same formation was exposed in the crater. No doubt, Shoemaker thought, the glass had been formed when the nearly pure quartz sandstone fused. For quartz to fuse in this manner, the temperature had to be about 1500 degrees celsius, some 300 degrees higher than the hottest lava flow. Not a collapsed salt dome, not even a volcanic explosion—Shoemaker now knew that only a high speed impact from an asteroid from space could have created enough heat for this glass to form. After almost half a century, Barringer at last was proved right: Meteor Crater was indeed an impact crater.

Shoemaker had to wait to test his idea of an impact origin for the crater until he'd completed his doctoral course and examination at Princeton. He had expected to finish his dissertation while continuing his uranium work on the Colorado–Utah border. By 1956 the roster of known uranium deposits was building quickly, and material for the growing nuclear power industry was known to be more abundant than expected, but now plutonium was the problem. This was the height of the cold war, the "balance of terror," as Churchill called it, between the United States and the Soviet Union. With the arms race in full swing by 1956, the amount of plutonium needed for the increasing numbers of weapons exceeded the supply. Enter Project MICE—for megaton ice-contained explosion. Simply stated, this involved wrapping a blanket of uranium around a nuclear device. The resulting explosion would breed a lot of plutonium. The explosion could occur buried in ice or in salt; however there was no body of ice, at least none under US control, large enough to handle an explosion equivalent to a million tons of dynamite that would produce the needed plutonium. So the answer came down to searching for large, relatively pure bodies of salt.

Shoemaker was asked to participate in MICE to answer a specific question: Would this underground explosion produce an eruption at the surface, somewhat analogous to the eruptions of ancient volcanoes in the Hopi Buttes? Shoemaker concluded that salt melted in the nuclear blast would not erupt at the surface at a later time.

But Shoemaker did find something else to worry about. Visiting the Nevada nuclear test site at Yucca Flats, he studied two 100-meter-wide craters, fancifully named Jangle U and Teapot S. Small subsurface 1.2-kiloton nuclear explosions had formed them—nothing too large but large enough to give an idea of what much larger forces could create. "I discovered that the shock-melted materials that had been close to the nuclear device were dispersed along the floors of both craters," Shoemaker remembers. The plutonium produced in the large explosion might not be neatly stowed in a puddle on the floor of the explosion cavity; instead it might be hidden in the larger volume of broken rock.

To extrapolate the effects that he observed in the small nuclear craters to those expected from a much larger explosion, Shoemaker decided to visit Meteor Crater. By this time he was almost certain that the crater was the result of an impact. "I was astonished to discover that the structure of Meteor Crater was pretty much a scaled up version of that of Teapot S," he remembers. But Meteor Crater taught him something else. Barringer's attempt to find a large iron meteorite under the crater was defeated because the meteorite, melted by the shock of impact, was largely dispersed in a thick lens of breccia, or broken rock, beneath the crater floor. Similarly the plutonium formed in a contained megaton nuclear explosion might be widely dispersed—both difficult to use and extremely dangerous.

Project MICE was abandoned later when other ways of increasing plutonium production were found. For Shoemaker the effort wasn't wasted. The project gave him a detailed understanding of the structure of an impact crater and the mechanisms by which it is formed.

A VISIT TO THE CRATER

In 1991 some 200 scientists from all over the world assembled at Meteor Crater to see the final resting place of an asteroid that had come to stay. Some had traveled from the soon-to-be-former Soviet Union to explore this place. "Miss a trip down Meteor Crater with Gene Shoemaker?" someone said. "You gotta be kidding!" As we clambered down the three-quarter mile wide bowl, slithering along its slope, the immensity of what happened so long ago became clear. What was causing us so much sweat, cuts, and bruises was carved out by a rock from space.

Gene Shoemaker *is* Meteor Crater, for although others suggested that it is the remains of an impact, Shoemaker proved it. He discovered layers of sediment that had been turned upside down, the new mineral coesite that was formed by the shock of the blast, and a host of other geological clues. All this created a picture in his mind of a thunderous impact, a mushroom cloud, and a parcel of land 2.5 kilometers wide, including the crater and the surrounding rim, that permanently reshaped itself in 5 seconds.

The vision of the ancient impact Shoemaker described to us strengthened as we started down the crater. It looked huge. How could an asteroid no larger than 50 meters in diameter have done such a thing in 5 seconds? Among the scientists at the crater rim was Fred (dirty snowball) Whipple, who had suggested more than 40 years earlier that comets were conglomerates of ices and dust orbiting in the solar system. At 85 years of age, he was still one of the most active conference participants that week. But as he looked down the precipitous trail, he decided not to attempt the hike. "I've done it before," he announced. "This time I'll walk the perimeter." Whipple's plan was hardly a visit to a shopping mall—it meant several hours of hiking across steeply rolling rocky terrain—but he did it.

Shoemaker did not intend to start us down the crater right away, instead we crept across the steep talus slope several hundred feet below the rim, hugging the wall and avoiding loose stones displaced by other hikers. We visited several interesting outcrops of rock layers that the impact had turned upside down. Shoemaker showed us other evidence he had chronicled over the years that pointed to the direction, speed, and type of the asteroid's fall.

Suddenly—very suddenly—Marsden whooshed past me as everyone struggled down the talus. With a relieved grin, he slalomed to a stop inches from Shoemaker, who looked at him somewhat startled. "Gene! Just trying to keep up!" he quipped. One person descended on crutches, another in a suit, tie, and vest.

As I made my way to the center of the crater floor, I finally caught my breath. There was water there—warm water, but it tasted so good! I looked up at the surreal late afternoon scene. Surrounding me were the towering walls of the crater, all carved out in 5 seconds.

We started back up. Shoemaker's group seemed to be heading straight up the crater wall. I had had enough, so I took an easier way on an old mule trail. As people passed me, their conversations now contained less science and more politics. Back in that early summer of 1991, the world was in such upheaval. The middle of this crater was somehow an appropriate place for a visiting Russian astronomer to bemoan the changes taking place in his country: "Four whole generations have grown up on Leningrad," he said, wiping perspiration. "Is it really necessary to toss out such a tradition and return to St. Petersburg?"

Finally making it to the top, I looked back at the huge crater now lit by a setting sun. Exhausted, we sat down to a rimside dinner, united in our appreciation of the importance of learning about these tiny marauders of the solar system. Despite the beauty and grandeur they created in sculpting the Earth, we appreciated that they could bring about untold damage.[4]

ℭ *10* 𝔇

A Turn on the Road to the Moon

N *ow strongly suspected to be the result of a hit from an asteroid from space, Meteor Crater sparked Shoemaker's curiosity. What could* hit the Earth from out there? What has hit the Earth? The answers would be hard to find on Earth, since most of the ancient craters have eroded or are buried, but on the nearby moon, a crater lasts almost forever. To answer his questions, Shoemaker needed to study the moon.

In 1956 Shoemaker spoke with Thomas Nolan, then director of the U.S. Geological Survey, about the possibility of launching a photographic study of the moon to produce the first geologic map ever made of something other than the Earth. It may have sounded wild at the time, and Nolan was the first person to hear the details of his moon dream, but to Shoemaker's surprise, Nolan didn't laugh. "It is to his credit," Shoemaker remembers, "that Nolan took the project quite seriously." He steered the young geologist to the right people, who explained that while a topographic map of the moon had been thought of, a map emphasizing its geological features had not. A year later the moon-mapping project was still at the idea stage, kept pretty much in the background as other work, particularly the MICE project, demanded Shoemaker's attention.

123

The early days of 1957 were not the time to sell the government or the military establishment on space exploration. Large missiles were being built and tested at places like White Sands in New Mexico by people with big ideas, like Werner von Braun, once a rocket scientist from Nazi Germany, now working for the United States. But when these same people suggested that the rockets being tested—particularly the Redstone and the Atlas—could some day hurl satellites into orbit around the Earth, and maybe on to the moon, they were ordered to forget the idea. Missiles were for defense, they were told, not for scientific exploration. Mr. Wizard, the popular science teacher, held no fascination for the U.S. government.

All that changed on October 4, 1957. Returning from a MICE meeting to a survey camp in the Hopi Buttes, Shoemaker heard big news on the portable radio. The Soviets had launched an artificial satellite called *Sputnik.* "Oh hell!" he winced, "I'm not ready for that yet!" But if Shoemaker wasn't ready, neither was anyone else in the nation. At the highest levels of government, U.S. policy turned 180 degrees overnight. The United States couldn't let its archenemy take control of space by default. Now there had to be a satellite program. Although the Naval Research Lab had its Vanguard satellite program, its first attempt collapsed in a cloud of exploding metal on the launch pad. Von Braun was put to work to get something—anything—into orbit within 90 days. Barely making the deadline, in 1958 a *Jupiter-C* rocket launched *Explorer I* into orbit. People cheered across the country, and the Pentagon brass breathed a sigh of relief. There would be more science in the schools. Where before Shoemaker worried that people would scoff at his ideas, now he was being listened to.

After MICE was shut down, the Shoemakers moved to Menlo Park on the San Francisco peninsula. With a group of people interested in exploring the moon and planets, Shoemaker participated in a colloquium at the start of 1959. He then began making a precursory lunar map to demonstrate that one could explain the stratigraphy of rocks exposed on the lunar surface.

BACK TO METEOR CRATER

With the National Aeronautics and Space Administration (NASA) established under the guidance of Senator Lyndon Johnson in 1958, Shoemaker believed his road to the moon was about to be paved. But despite his attempts to get the U.S. Geological Survey involved in a moon program, Shoemaker kept running into indifference. Then in the spring of 1960, something that seemed innocuous—an article in the *Journal of Geophysical Research* by geologist Joseph Boyd—set Shoemaker on yet another turn on the road. The article established temperature and pressure conditions under which a high pressure form of silica may crystallize. The impact of a comet or an asteroid, Shoemaker thought, would create an environment for such a mineral to form in the real world. Could this be the signature, the giveaway for other impact craters? With colleague Beth Madsen, Shoemaker intended to look for this mineral in the extensive collection of shocked rocks he had gathered from Meteor Crater. But before he had a chance to do that, another colleague, Edward Chao in Washington, detected the mineral in a specimen from Meteor Crater on display at the National Museum.

Soon Shoemaker and Madsen found samples of the mineral and mapped out the occurrence of this high-pressure mineral at Meteor Crater. This newly identified mineral was named coesite, after Loring Coes, the physical chemist who had discovered its crystalline phase in the laboratory. Thus, by 1960 there was a reliable basis for demonstrating that other craterlike structures on Earth were the wounds of ancient impacts. The timing was perfect. The fledgling NASA finally decided to fund Shoemaker's proposal for a detailed quadrangle-by-quadrangle geologic map of the crater-pocked moon.

The summer of 1960 saw another turn on the road to the moon. The Shoemakers planned a trip to the International Geological Congress, which was being held that year in Copenhagen. On the way they planned to vacation in southern Germany and visit the Ries basin, a 27-kilometer-wide structure that Shoemaker

suspected of being an impact crater. "My father died that spring,"
Shoemaker noted,

> and so we encouraged mother to join us. We bought a VW
> microbus, took delivery of it in Hamburg, headed south along
> the Rhine to look at some volcanic craters, then turned east
> toward the basin. We drove through rain and reached the site
> near the end of the day.
>
> The rain had stopped. By the light of the setting sun we looked
> at the shock-formed rocks that had been supposed to be vol-
> canic. I took one look at these rocks with a hand lens. No
> question these were impact rocks!
>
> With night falling, the three found a wooded campsite. The
> next morning they went to the post office, airmailed two sam-
> ples to Edward Chao in Washington, and then visited the St.
> George's Cathedral in Nördlingen. "The whole cathedral was
> built of suevite—it was made of stuff that had suffered the
> shock of the impact," Shoemaker marvelled. "And at that
> moment, I was the only geologist in the world who knew it![1]

Back in Washington Chao's X-ray diffractometer quickly confirmed Shoemaker's suspicion that an impact some 15 million years
ago had left its coesite signature. Whatever hit back then must
have been pretty big, something perhaps 2 kilometers across. The
discovery sparked a flurry of activity, both among German geologists, who had a field day with their newly anointed impact crater,
and among other geologists, who were interested in finding other
impact sites around the world. In Canada, an aerial search for
impact craters had already led to the discovery of some promising
sites, including a 3-kilometer-wide crater a short distance from
Kingston, Ontario, that formed when something slammed into the
Earth 500 million years ago, one of the oldest events on record.

Years later Shoemaker revisited Nördlingen, this time to receive, along with Chao and a German geologist, the first Ries cul-
tural prize. "It was so good to see that geology is part of their
culture," Shoemaker noted.

A GEOLOGIC MAP OF THE MOON

Now that the lunar mapping project was started, Shoemaker proceeded in much the same way he did out on the Colorado plateau. He wasn't on the moon—not yet—but he had pictures. Some 40 years earlier, on exceptionally steady nights when the 100-inch Hooker telescope at Mt. Wilson Observatory had been new, Francis G. Pease had taken such high-quality photographs of the moon that they picked up craters as small as 1 kilometer in diameter. Shoemaker had enlargements made of the region around the crater Copernicus, a huge feature that might have resulted from a comet impact about a billion years ago. From these photographs his team made the first geologic map of a lunar feature. Besides Copernicus the map showed a whole set of geological features in a region about the size of Arizona. Later Shoemaker expanded the mapping into adjacent areas. Once completed, the project included 50 quadrangles covering much of the entire side of the moon visible from Earth.

In 1960 Shoemaker joined the television team for Project Ranger, the widely publicized program to send a small spacecraft to the moon. The next year the U.S. Geological Survey set up a formal Branch of Astrogeology with Shoemaker as branch chief. Although Shoemaker didn't want to administer the branch, that seemed to be the best way to advance the program. Despite President Kennedy's decision in May 1961 to send an American to the moon by 1970, the concept of geologic mapping as a technique for deciphering the history of a planetary body was unfamiliar to most of the managers at NASA, and it was still difficult to get funding for lunar research. Still the program expanded slowly. With two of the highest resolution telescopes then available, the 36-inch diameter refractor at Lick Observatory in northern California and the 24-inch refractor in Flagstaff, Arizona, Shoemaker's team worked to decipher the stratigraphy and geologic history of regions that were destined to be explored by unmanned spacecraft and later by *Apollo.*

The team showed that a covering of material (called an ejecta blanket) thrown off when something hits the moon lies on top of much older blankets and other geologic structures. These blankets gave the team a way of ascertaining the relative age of the craters. The blankets also showed that when a comet or asteroid hits the moon, pieces of the lunar surface are hurled up and land again, forming secondary craters that could also help establish the sequence of events and the stratigraphic relation of ejecta blankets. It was a good start, but Shoemaker felt he needed more: a special telescope dedicated to lunar mapping. Once he received funding for this 31-inch diameter telescope, Shoemaker decided to move his family and the headquarters of the U.S. Geological Survey's astrogeology branch to Flagstaff, Arizona.

By then the Shoemakers had two daughters and a son. Flagstaff wasn't just close to Meteor Crater, it was also a good place to raise a family. But it would no longer be a turn on Shoemaker's road to the moon. The medical requirements for astronauts were unyielding—if you had sniffled once or twice in your life, you could forget it. In 1963 Shoemaker was diagnosed with Addison's disease. Although it was under good medical control, Shoemaker knew that he could not be an *Apollo* astronaut. If he planned to go to the moon at all, he would have to do it on Earth.

In 1964 Harry Hess, an old friend and professor from Princeton, chaired the committee that would review the scientific qualifications of scientists who wished to join the *Apollo* astronaut corps. "What I thought about in 1948 seemed to be coming to pass," Shoemaker says.

> Harry Hess chose to head up the committee himself; he asked several people, including me, to work with him. But when the time came for the committee to do its work, he was on sabbatical in England so he said, "Here, Shoemaker, you chair the committee!" Instead of being there at the head of the line with my application to go to the moon, I ended up chairing the committee that reviewed the other applicants.[2]

Out of a thousand applications for scientist astronauts, only 15 qualified. Of those, only six passed NASA's incredibly tough physical exam. And only one—Harrison Schmidt—actually made it to the moon.

Exploring Craters
From Ranger to Apollo

Now there was a bright new moon, and as usual in that phase its horns were tilted towards the east. Suddenly the upper horn split in two. From the midpoint of the division a flaming torch sprang up, spewing out, over a considerable distance, fire, hot coals, and sparks. Meanwhile the body of the moon which was below writhed, as it were, in anxiety, and to put it in the words of those who reported it to me and saw it with their own eyes, the moon throbbed like a wounded snake. Afterwards it resumed its proper state. The phenomenon was repeated a dozen times or more, the flame assuming various twisting shapes at random and then returning to normal. Then after these transformations the moon from horn to horn, that is along its whole length, took on a blackish appearance.[1]

*F*orgotten for almost 800 years, this narrative by a group of British monks captured the fancy of a geologist named Jack B. Hartung in 1976. He suspected that this report could be an accurate observation of a comet or an asteroid hitting the moon.[2] The chronicle describes an extraordinary event—the upper part of the crescent splitting in two as debris from the crash blasts a dark cloud hundreds of miles across the moon and finally a pall of dust giving the moon a short-lived atmosphere and preventing its light from reaching us—blackening it in a sense.

But if this were a major impact, where is the crater? On the moon, where there is virtually no erosion, there is little to indicate relative youth. Its surface is pockmarked with craters that do not disappear. New craters even form on top of old craters. However there is a key: As debris flies across the moon after an impact, it lands again in long streaks of small craters and boulders that are easily seen from Earth as systems of rays. Tycho, the largest crater to be dug out of the moon in relatively recent times—it is thought to be only a hundred million years old—is the best example. This crater is named after the famous sixteenth-century Danish astronomer who was also familiar with impacts in a sense—he lost part of his nose in a fight. Although the moon's craters become virtually permanent features, the youngest craters are surrounded by long, soft rays of rocky material that radiate from their centers. These systems of rays do not last more than a few hundred million years. Thus if a crater has its ray system, it is probably relatively young.

Tycho's ray system stretches out over half the moon's surface. The moon's second most substantial system of rays originates in a crater on the far side, never visible from Earth. This crater is named Giordano Bruno in memory of the scholar—a contemporary of Galileo—who was put to death in 1600 for suggesting that Earth was but one of many worlds and these worlds could harbor intelligent life. Crater Giordano Bruno is just beyond the side of the moon that is visible from Earth, but it has been photographed by spacecraft. The crater is about 20 kilometers in diameter, the size of a small city. Although it is only one-fifth the size of Tycho, its ray system is almost as large. Had the object that caused it, probably about 2 kilometers across, fallen on Earth instead, the result would have been catastrophic, perhaps even capable of wiping out civilization.

Even with the evidence of Giordano Bruno's youthful ray system, however, we can't say conclusively that the crater was formed in 1178. The chronicle is dated the end of June, placing it near the maximum of the Taurid meteor shower, which consists of a large number of remnants of a comet that must have disintegrated long ago. Comet Encke is the largest member of the Taurid swarm.

Illustration of rays on the moon (see Chapter 11). The author took these photographs during a partial eclipse of the moon. In the brighter photograph, the large crater Tycho is near the top, and its ray system, stretching away from it and extending across the moon's surface, can be seen in both pictures. During eclipses the long rays of material coming from craters like Tycho stand out more clearly.

Kenneth Brecher of Boston University suggests that a smaller relation of Encke's comet could have struck the moon as it passed through the Taurid swarm in 1178.[3] Other scientists suggest that the same effects could have been produced by a meteor in our own atmosphere chancing across the position of the moon. But some unusual phenomenon in the Earth's own atmosphere may also have produced a similar effect. In any event, by far the weakest link in this chain of evidence is the ancient observation itself, since no one else recorded the event. Bradley Schaefer, who studies the visibility of thin lunar crescents, insists that the moon was too low on the night mentioned in the chronicle, which is June 25 by our calendar.[4] However it would not matter much if the date were off by one and the event really took place on June 26. Was a bright comet visible that year that could have hit the moon? (There was, but its appearance 6 months early rules it out as the possible culprit.) Or was the object an asteroid, which would not have been observed? According to physicist Graeme Waddington, who suggested the June 26 date after studying the *Chronicles,*

> Gervase is highly regarded and the details in his Chronicle are considered to be remarkably accurate. It is precisely because of this high esteem that Gervase's record of the "Canterbury event" must be taken seriously. . . .[5]

Unless someday we can collect material from Giordano Bruno, we may never know whether it really is as young as some evidence might suggest.

TO THE MOON WITH RANGER

Eight hundred years after the British monks looked toward the moon in shock and wonder, other eyes looked in wonder at the moon. By 1960 it was time to explore the lunar surface with spacecraft. Although the crater Giordano Bruno mystery was not a factor in the late 1950s when lunar exploration began—the crater's existence wasn't even known until the Soviet unmanned spacecraft *Luna III* passed by the moon in October 1959—the mystery of how so many craters were formed on the moon was a prime motive for going. The Earth and the moon were likely formed at the same distance from the sun, but the cratering record on our active planet has been largely erased by erosion and other geological and climatological processes; not so on the moon. If we could understand what happened there, we would understand what happened here. It is that logic that drove Shoemaker to the moon—if not in person, then through the eyes of machine probes and other men.

Shoemaker was asked to join the *Ranger* television team around 1960, along with Gerard Kuiper and Harold Urey, then two of the leading authorities on the moon. Shoemaker was being let in on the ground floor of Project *Ranger*, an exciting new endeavor, and he accepted eagerly. Conceived by a group of engineers at the Jet Propulsion Laboratory (JPL), *Ranger's* aim was ambitious almost beyond belief for its time. "The Jet Propulsion Lab was a really vigorous place," Shoemaker notes. "They were going to build a spacecraft fast and send it to the moon." The hexagon-shaped spacecraft bristled with an array of instruments for conducting several experiments. There was a television camera, and a gamma ray spectrometer to measure detectable radioactive elements on the moon. Tucked inside was a balsa instrument container atop a retrorocket. The idea is that the balsa package would detach from the main spacecraft as it approached the moon. The retrorocket would fire as long as it could, slowing down the balsa container and landing it on the surface. Then a small seismometer would right itself while a small antenna pierced the encasement and started sending data to Earth.

After the first two test *Rangers* were successful in trial runs, all was ready for the real mission. But then the problems started. *Ranger 3* missed the moon by almost 40,000 kilometers a few days after its launch on January 26, 1962. A few months later, *Ranger 4's Atlas-Agena* rocket worked flawlessly, sending the craft on a direct course to the moon. But a failure on the spacecraft prevented any of the experiments or cameras from operating, and the craft, its balsa container still aboard, shattered on the moon. *Ranger 5*, launched in the early fall of 1962, came the closest to success of the original *Rangers*, but despite the fact that all its experiments seemed to work, it missed the moon by a few hundred kilometers.

By this time, the hope and excitement at the JPL had turned to gloom. "Either the spacecraft was great," Shoemaker lamented, "or the *Atlas* booster was great. Trouble is in five tries, we couldn't get a good rocket and a good spacecraft at the same time."[6]

After *Ranger 5's* dismal failure, NASA put the whole program on hold while the project was redesigned. There was another consideration. In May 1961 President Kennedy had committed the United States to landing a man on the Moon by the end of 1969, so engineering firms all over the nation were gearing up for mighty Project *Apollo*. After the failure of *Ranger 5*, scientists still had no idea what the lunar surface was really like. Would it be firm enough to support a manned spacecraft? Where was the best and safest site to land a manned spacecraft? Thinking that the problem with *Ranger* was that the craft design was too complicated, NASA tried simplifying the idea on the notion that a few good close-up pictures would be better than a spacecraft full of expensive but useless instruments. The balsa container vanished as well as the boom with the gamma-ray detector. Instead a simple conical tower appeared on top of the hexagonal frame to support a set of six television cameras. Of all the scientific teams associated with the Ranger project, only the Kuiper–Urey–Shoemaker team directly benefitted from the new design. A good picture was all that mattered now.

Although this sounds unscientific, the real cause of the early *Ranger* flops was simply bad luck. After all a similar version of the same spacecraft flew successfully to Venus as *Mariner* 2 in 1962. In any event, four newly designed spacecraft were built. Finally at the end of January 1964, all was ready for the launch of *Ranger* 6. Shoemaker recalls the cutting tension at the JPL:

> The *Atlas* performed like a jewel. The *Agena* performed like a jewel. The spacecraft was fine. It was sending a clear signal as it made its final approach to the moon. Everyone in the lab

was huddling around the speakers, listening to the reports from the Goldstone tracking station. When the spacecraft was about a thousand kilometers from the lunar surface, the final critical signal was sent to turn on the cameras.[7]

Nothing. "No video." The signal was sent a second time. "Still no video." For the next few minutes the room was silent except for the announcer's repeated, "Still no video." Then the spacecraft crashed into the moon, its faint signal silenced. The control room at JPL was as silent as a crypt.

The cause of *Ranger 6's* frustrating failure turned-up soon afterward—a simple short circuit had killed the television cameras during the strain of launch. When it was time for *Ranger 7* that summer, once again the launch went flawlessly, and as *Ranger 7* approached the moon on target a few days later, the control room at JPL was filled with people. As the spacecraft neared its thousand-kilometer distance from which it would begin photography of the moon, everyone in that room knew that JPL's future rested on what would happen in the next few seconds.

The signal was sent to turn on the cameras. "We have video!" It was an indelible moment. "Suddenly everyone was jumping up and down," Shoemaker raved. "We didn't even know what the pictures were yet, but that didn't matter. There were pictures!"

Before the craft smashed into Mare Cognitum, its cameras caught the first view of what the ray of a crater looked like up close. It recorded a beautiful tapestry left over from the impact of the comet or asteroid that produced the Crater Tycho a hundred million years ago. The crash point was near one of the bright rays that extend from the crater—rays filled with hundreds of secondary craters that were produced when material hurled upward from Tycho's main impact point landed far away. Some of these craters were as large as 100 meters or as small as 2 meters across.

The JPL desperately needed the success of *Ranger 7*. For *Ranger 8*, the scientific team tried to persuade engineers to aim the spacecraft closer to the moon's terminator, the moving zone of lunar sunrise or sunset where lower sun angles would result in far better pictures. However the engineers feared that these areas would not

provide sufficient light for the craft's cameras. It was not until
Ranger 9's assault on the crater Alphonsus in March 1965 that the
program reached its zenith. *Ranger 9* was aimed at one of the
moon's most interesting areas. In 1958 the Soviet astronomer Ni-
kolai Kozyrev had seen whitish glows in this crater, and his observations through a spectroscope showed evidence that gases were
being emitted. *Ranger 9* was sent to answer a specific question:
Was there volcanic activity in Alphonsus? *Ranger 9* arrived just
after Alphonsus' sunrise, snapping almost 6000 pictures before
crashing into the crater. The pictures revealed some halo craters
that might be volcanic in origin. Apparently there was plenty of
past volcanic activity there, but no current action.[8] Like the earlier
Rangers, *Ranger 9* taught scientists about the fine structure of the
lunar surface.

ON THE MOON WITH SURVEYOR

Soon it was time to make the transition to a craft that would
do everything the earlier *Rangers* tried to do and more. Instead of
having a wooden box bounce around on the moon, *Surveyor's* entire
Christmas-tree-like frame would land so delicately that it would
barely disturb the soil it lighted on. It would then photograph the
surface for more than a week before setting down to rest for the
long and frigid lunar night during which temperatures would
plummet to more than 200 degrees below zero. Shoemaker was
invited by JPL to participate as an investigator.

The early designs for *Surveyor* loaded the craft with cameras
and other instruments, but as failure after failure dogged the *Ranger*
program, NASA and JPL decided to scale down *Surveyor*. "They
threw off all the instruments," Shoemaker laments, "leaving only
a single camera atop the tree as an engineering test."

Wondering when he would ever get his first close-up view of
the lunar surface, Shoemaker spent much of his time experimenting
and planning how the camera would do its work once it finally
landed. But the first view happened sooner than he expected. In

early February 1966, the Soviet craft *Luna 9* gently landed on the moon much as the original *Rangers* had tried to do. But Western nations could spy on this craft, thanks to Sir Bernard Lovell, England's famous radio astronomer, and the giant dish of the world's largest radio telescope at Jodrell Bank. As they examined *Luna 9's* signal, someone noticed that it resembled transmissions that wire services used for newspaper pictures. Wondering if it were really as simple as that, the observers rushed to borrow a facsimile machine and played the signal through it. It worked! They secured a good, though somewhat distorted, picture that showed a surface strewn with rocky rubble.

With memories of the *Ranger* experience still fresh in their minds, Shoemaker's team did not expect anything from the first *Surveyor*. "Of all the people on the *Surveyor* project," Shoemaker says with not too much exaggeration, "only three of the engineers were confident that the craft would actually make it to the Moon. They had really worked that spacecraft over." So when *Surveyor 1* finally made it off the launch pad atop an *Atlas–Centaur* rocket on May 30, 1966, a few months after the Soviet venture, Shoemaker saw it as only the first step in a long process where a single malfunction would doom the journey. At least, most people thought, *Surveyor 1* would be a good learning experience. There was a long way to go, and getting past the first stage of the *Atlas* was the easiest step. Next came the *Centaur's* firing—the upper stage designed to hurl the craft into Earth orbit and then to refire, sending it off to the 25,000 miles per hour velocity needed to escape the bonds of Earth and head off to the moon. Although the *Centaur* had succeeded in tests, this was its first operational flight.

The *Centaur* did its job. Now cautiously optimistic, Shoemaker and his team watched the flight phase from the Earth's orbit to the vicinity of the moon go well, and they assembled at the control room at JPL with the craft still operating nominally. Traveling at 6000 miles per hour only 60 miles away from the moon, *Surveyor's* main rocket fired for about 40 seconds, slowing it down to 250 miles per hour as a radar altimeter connected to an on-board flight programmer monitored the quickly approaching moon. Everything was still fine as the retrorocket dropped away and three smaller rockets took over, slowing the craft to a virtual stop 13 feet above the surface. The rockets then shut off to leave the surface below as pristine as possible. *Surveyor 1* fell quietly to a landing, the strong shock absorbers in its landing legs absorbing much of the shock.

Back in the control room on Earth, the engineers were stunned. "It really landed! We were still communicating with it!" Shoemaker reveled. "It was the most surprised bunch of people you ever saw!"

The next hour passed interminably as Shoemaker and his team waited for the spacecraft's systems to be checked out. But then the camera was on and taking pictures. "I slept about 2 hours over the next 5 days," Shoemaker remembers of the hectic week spent looking at the stream of data from the camera. In that newly finished Spaceflight Operations Facility at JPL, Shoemaker had attained his dream. At *Surveyor's* camp there was a blanket of material that had been thrown off by ancient impacts. The surface was riddled by impact craters from a few centimeters to 30 meters in size.

The second *Surveyor* did not make it to the moon intact, but in April 1967, the third one landed right in the middle of a crater several tens of meters wide. Although the landing was a fine place from which to study what a crater floor was like, the camera couldn't see beyond the crater walls. The fourth *Surveyor* failed, and were it not for a special aspect of its design, the fifth one, launched in September 1967, would have flopped as well. After it turned off its rockets 13 feet above the surface, *Surveyor 5* settled on the wall of a small crater not much larger than it was. Then it started to slide into the crater, its feet gouging out trenches for about a meter before the craft finally came to a stop perched at an angle of almost 20 degrees. Fortunately the spacecraft was designed not to tip over if it set down at an angle that steep. Of course nobody knew of *Surveyor's* dangerous adventure until it was over and the camera was turned on and looking about, taking in the crazy angle and the three footpad trenches.

Launched at the end of 1967, the sixth *Surveyor* took the first stereographic pictures of its site. After its main photographing mission was done, the rockets were turned on and the craft hopped a few meters to a new location. By shooting the same areas a second time, Shoemaker's team put two pictures together to see a three-dimensional representation of what the surface looked like. The stereographic pictures provided a sense of depth that would be very useful when working out distances and heights of the surrounding terrain.

The team had hoped *Surveyor 6* could be targeted to a highland area, since the lunar plains, or maria, had been pretty carefully studied by that time. But the idea of sending *Surveyor 6* to the region of Fra Mauro was rejected; instead *Surveyor 6* went to the Moon's belly button—Shoemaker's term for the Sinus Medii, the small lava plain right at the middle of the side of the moon pointed toward Earth.

With just one *Surveyor* to go, Shoemaker and the other scientists had a plan. "Let's see something different," they suggested. "Let's go to Tycho!" By this time another series of missions called *Lunar Orbiter* had shown the Tycho area to be very rough, so the idea was risky. But the payoff would be handsome. Some 90 kilometers wide, with walls 20 kilometers high, Tycho is most probably, Shoemaker suggests, the site of a comet nucleus impact that happened perhaps a hundred million years ago. At that time dinosaurs were in the middle of their dominance here on Earth. Had these creatures had the presence of mind to watch the sky, they might have seen a comet getting brighter and then suddenly vanishing in a flash of light as it crashed into the moon's southern hemisphere. The crash raised a moonwide cloud of dust as huge chunks of material tore out of the surface and flew as far as halfway around the moon before crashing again to form secondary craters—something quite like what the British monks thought they saw in 1178. We can still see how this impact affected the moon, using a small telescope to view both the crater and its system of rays. The rays are seen at their best when the moon is near full phase.

To Shoemaker's surprise, NASA welcomed the proposal despite the fact that Tycho was far from any of the proposed *Apollo* landing sites. *Surveyor 7* landed quietly on a gentle 5-degree slope on Tycho's northern rim. The first picture told how precarious the landing was: *Surveyor 7* was in the middle of a field of boulders, many of which could have punctured the craft or tipped it over. There was a great variety of rock types, including rocks crushed and deformed by high pressure. "Tycho offered a much younger surface than anything else we had seen," Shoemaker noted.

Surveyor's call on Tycho happened a decade before the discovery of a worldwide layer of iridium led nobel laureate Luis Alvarez, his son Walter, and their colleagues to suggest that a comet or asteroid impact had destroyed most of the life on Earth 65 million years ago. And a quarter-century after *Surveyor's* work was finished, the remains of a larger crater than Tycho were found at Chicxulub on the Yucatan Peninsula of eastern Mexico. Chicxulub and Tycho would appear to have little in common. While Chicxulub is buried, Tycho and its moonwide system of rays dominate the view through even a small telescope. But comet impacts possibly caused both features. Chicxulub is at least twice Tycho's diameter but possibly only two-thirds as old.

APOLLO: THE LAST STEP

Shoemaker was the principal investigator for the field geology experiments for the first three *Apollo* moon landings. With a great deal of enthusiasm, his team set up a mock lunar-landing site, complete with small artificially made craters and a full-size model of the lunar module to give practicing astronauts a chance to practice geological field techniques. "Some of those test pilots were very good observers," Shoemaker remembers, adding that their flight training and alertness gave them the potential to be ideal field geologists—if they could get sufficient training.

Since hanging either tools or devices on an astronaut, such as a geologist's hammer, was out of the question, Shoemaker came up with an ingenious device he called Jacob's staff that the astronaut could carry like a cane. It would have a TV camera, an antenna, and a transponder, so that the lunar module could track it. All the astronaut would have to do was pull a trigger. Then the staff would take a picture while its leveling devices recorded the orientation. But NASA never funded the staff, and Shoemaker worried that the astronauts were not given enough time to practice the methods of field geology. Would *Apollo* be a more complex version of the Mercury "man in a can" program where the astronaut went along for the ride while ground controllers and simple computers gave all the instructions? As a field geologist, Shoemaker hoped for more than that from the *Apollo* moon walks.

On Christmas Eve 1968, a few months after the last *Surveyor* had seen Tycho, three astronauts were orbiting the moon and sending an unforgettable Christmas greeting to all of us on "the good Earth." It was a poetic and moving way of building excitement for the landing that was to follow a few months later. On July 16, 1969 *Apollo 11* surged to life and roared off its launching pad. This was the culmination of a generation's dream. As Armstrong, Aldrin, and Collins left the Earth's orbit and headed moonward, the flight was proceeding as though manned landings on the moon took place every day. But at the very last minute, the astronauts saw trouble on the moon. "We had carefully mapped every crater in the landing strip on Mare Tranquilitatis," Shoemaker recalled. It was an ellipse about 10 kilometers and 2 kilometers wide. But although *Lunar Orbiters* 4 and 5 had recorded some gravity anomalies in the moon—the mascons, or mass concentrations associated with the moon's huge lava-filled basins—the *Apollo* planners had ignored them. As a result, instead of heading for the center of the ellipse, *Apollo 11* was about to land near the far end—right down in one of the largest craters in the whole ellipse.

"Every *Surveyor* that made it to the moon—blind—landed safely," Shoemaker notes. "But the first time we had a man in the spacecraft we really needed him." The crater's blanket of ejecta was full of huge boulders that would have caused the *Eagle* to tip over and crash. As Neil Armstrong urgently steered the craft away from the crater, Buzz Aldrin noted his progress, both downward and across:

Houston: 60 seconds [of fuel left]
Apollo: Altitude 1600
1400
400 feet, 8 forward
300 feet, 47 forward
13 forward
11 forward
down one-half; picking up some dust . . .
4 forward, drifting to the right a little
Houston, urgently: 30 seconds!
Apollo: 4 forward, 40 forward[9]

With less than half a minute of fuel left, Armstrong set *Eagle* between two of the crater's rays. "OK, engines are off." One-sixth of the world's population heard Armstrong's next words: "Tranquility base here. The *Eagle* has landed."

Although Armstrong had by far the shortest of the moon walks, his 90-minute traverse was, to Shoemaker's mind, one of the most successful in the entire *Apollo* program. "He saw more stuff, and he made more pertinent observations, in the precious little time he had on the surface, than many of the astronauts who followed him."

For Shoemaker, *Apollo 11's* return to Earth was the beginning of one of his biggest disappointments. "NASA wanted to plot out every minute of every field traverse," he huffed. "That plan mitigates against discovery." Thinking as a field geologist, Shoemaker believed that the *Apollo* program, successful as it was, was a big lost opportunity.

Discovery begins when you see something you don't expect. Then you stop and say, "wait a minute, what is this? Maybe I'd better look over here." Then you start out on a completely unplanned traverse to understand what you have seen. *Apollo* never provided that opportunity.[10]

Harrison Schmidt, the lone geologist chosen for the astronaut program from the committee Shoemaker chaired 5 years earlier, managed to make some fascinating discoveries despite the confining schedule. The orange soil found in Shorty crater turned out to be an interesting glass formed in a fountain of ancient volcanic activity. But although Schmidt collected some, mission planners didn't allow the time to take the kind of care that was needed to see how the glass was distributed. *"Apollo* provided outstanding science," Shoemaker concludes,

> But the scientific discovery came from the analysis of the samples brought back, not from the observations of the astronauts. And that was the issue. Why send human beings into space, if not to be prepared for the unexpected?[11]

Shoemaker's condemnation of the *Apollo* program may be a little harsh; after all, this was hardly a normal geologic traverse being monitored from 240,000 miles away. Although the astronauts did not spend so much time at Shorty crater as a field geologist would have liked, Schmidt handled the situation expertly. As soon as his partner reported something highly unusual—the orange soil—Schmidt immediately warned him to delay collecting so that he could study the material in its original, undisturbed setting before he collected any samples. "Don't move it till I see it!" he said repeatedly. Only after Schmidt had a chance to inspect the soil did the astronauts start collecting samples, and the cameras recorded the scene in detail. Meantime a hot discussion took place on the ground as to whether it would be possible to extend the visit to Shorty crater. It was decided that extending the visit so far from the Lunar Excursion Module (LEM) would be too risky.

"My dream for *Apollo*," Shoemaker remembers,

> was to try to create the opportunity to show what a well-trained human being could do, what kind of science he could do on the spot. This is not a kind of science that most practicing scientists understand because they don't do it. But in the six Moon landings, we never demonstrated that important discoveries could be made from field observation.
>
> *Apollo* was not a highlight in my career by any means. The main issue for me was not flags and footprints, but to show why you needed a human being there.[12]

The proof of *Apollo's* failure in Shoemaker's mind comes from what happened afterward. "We stopped. Had *Apollo* really succeeded, we'd still be there exploring." We came, and we saw, but instead of conquering, we went home and did not return.

One important thing that the *Apollo* astronauts did, however, brings us back to the story of crater Giordano Bruno. On *Apollo 11*, Armstrong and Aldrin installed a small reflector off which Earth-based telescopes could bounce laser beams, thereby yielding very accurate measurements of any delicate wobbling motions of the moon, movements we call librations. With the *Apollo 14* and *15* missions as well as the Soviet *Luna 21* expanding the reflector network to cross the moon, astronomers Odile Calame and Derral Mulholland used McDonald Observatory's 107-inch reflector to fire laser beams at these reflectors more than 2000 times. In 1978 they independently calculated the possible effects of an impact exactly 800 years earlier and came up with the same conclusion Hartung did. The event, they wrote, "would have been not only visible but sufficiently apocalyptic to have justified the description given in the Canterbury chronicle."[13] Even more encouraging—though still not definitive—were their laser ranging results, which showed that the moon is indeed swaying by a few meters in a 3-year cycle. Like a huge bell vibrating after it has been clanged, the moon is acting as if it had been struck by a large object within the last millennium.

The *Apollo* astronauts also left a series of seismometers designed to detect moonquakes and impacts on the moon. In late June 1975 near the anniversary of the supposed Canterbury observation, the seismometers recorded the impact of 29-kilogram-sized objects.[14] Could these be part of the swarm that included the Canterbury object if it were real?

Of all the threads that weave this tapestry together, the laser-ranging results from Calame and Mulholland are by far the most impressive, clearly suggesting that the moon has been banged about in the recent past, possibly during the last millennium. It is my hope that some future astronaut may land near crater Giordano

Bruno and conduct field traverses that Shoemaker would be proud of—and sample materials not just at the surface but below it as well. For like all geological mysteries, the answer to the mystery of crater Giordano Bruno is written in rocks that await our return to the moon.

☾ 12 ☽

A Pretty Good Moon for You, Shoemaker!

If 1969 were a time for exploring craters on the moon, it was also a time to start looking for other things in space—not just craters but the comets and asteroids that cause the craters. It turned out that Caltech's Palomar Observatory had a small 18-inch diameter Schmidt telescope that was not being used very much. Palomar is most famous for its mighty 200-inch reflecting telescope, but three other telescopes atop that mountain northeast of San Diego also do important work. As Palomar's first telescope, the 18-inch has a noble history. Astronomer Fritz Zwicky used it to photograph fields of distant galaxies, and in these galaxies he discovered many exploding stars called supernovae. The 18-inch telescope is a photographic telescope capable of taking pictures of large areas of the sky at once—each film covers 8.75 degrees of sky, the equivalent of more than 17 full moons lined up. On a single film, a searcher could record a large nugget of sky. Here was an ideal telescope for searching for comets and asteroids.

The first thing that comes to mind when we think of asteroids is the large belt of them between Mars and Jupiter. We now know of thousands of those main-belt objects. But for an asteroid to hit the Earth or the moon, it has to wander way outside the main belt. It is these errant asteroids that Shoemaker wanted most to study.

A small subset of the Mars-crossing asteroids, objects we call Amors, go so far inside Mars's path that they approach the orbit of the Earth and briefly make close passes to our planet. Still other asteroids, called Apollos, actually venture across the Earth's orbit. In 1969 astronomers knew of about a dozen of these Apollo-type asteroids, each a kilometer or more in diameter.

THE EARTH AS A DARTBOARD

One of the most dramatic Apollos sprinted within 800,000 kilometers of Earth, about twice the distance of the moon, on the night before Halloween 1937. It has not been seen since, but that asteroid, named Hermes, will be back some day in the distant future, either to pass close by the Earth again or even to hit us. Meteor Crater is the result of what happened when an asteroid about 50 meters across—much smaller than Hermes—hit home.

If the Apollo asteroids pose a danger to the Earth, then the two most urgent questions are how many Apollos are there and more important, how close will they get to us? Becoming interested in that question in fall 1969, Shoemaker hired a Caltech scientist named Eleanor Helin "to track down every scrap of information" on the dozen or so Apollo asteroids that had been discovered up to that time. This included getting copies of the original discovery plates from places as far apart as Germany, Belgium, France, and South Africa to learn as much as possible about where the objects were in the sky when they were first spotted and how large they are.

For Shoemaker the problem of how to calculate the current rate of collision of objects with the Earth was already answered in the lunar craters. The crater record there however did not answer questions about the nature of the objects that hit—were they asteroids or comets? The 18-inch telescope, available as a part of Caltech, presented a golden opportunity.

With leftover funds from his role in the Apollo project, Shoemaker and Helin drew up a proposal in 1972 to take the 18-inch telescope on a test run to see what would be involved in searching for asteroids in the vicinity of the Earth. "Glo deserves a lot of credit," Shoemaker said, using Helin's nickname, "for pushing to start this program at the beginning." At the time Shoemaker thought that about 2000 asteroids the size of Hermes (2 kilometers or larger) could be around—all potential bullets aimed at Earth, each packing the wallop of a multimegaton bomb. He figured that if they were to photograph 250 fields a year, they would perhaps find some two new Apollos and two Amors each year. Their plan was to take a 20-minute exposure followed by a 10-minute follow-up shot, giving any new asteroid time to leave its signature as a trail of light across the star-filled photographs. "There was a lot of sweat and tears for each observing run," Helin recalls those difficult early days. And getting the telescope working was only part of the problem. "Roadblocks appeared on a regular basis," she explains. "I'm sure Palomar didn't think I would last a year." Her program is still going strong after 25 years.

Helin averaged seven films per night during her monthly observing runs at the 18-inch telescope. In summer 1973 only 6 months after the project started, she found her first object—one that was designated 1973 NA and now numbered 5496. Encouraged by the rapid find, they pressed on, with Helin doing most of the work. But that new Apollo was just a teaser; it was two-and-a-half long years before the next discovery. For a time after that, the finds remained scarce, although Helin discovered Aten, an asteroid orbiting the sun mostly inside the orbit of the Earth.

FINDING THE DARTS

By the program's fifth anniversary in 1978, Shoemaker knew that his estimate of 2000 near-Earth asteroids was about right but only if he included asteroids down to 1 kilometer instead of 2 kilometers in diameter; and the former were more difficult to find. The discovery rate was so discouraging that Shoemaker thought of applying for time on Palomar's larger Schmidt camera, the mighty 48-inch. Some 20 years earlier it had completed the Palomar Observatory Sky Survey, when it photographed the entire heavens in the northern hemisphere in more than 900 sections.

The first few years of the Shoemaker–Helin Planet-Crossing Asteroid Survey disappointed Shoemaker but they also set him to work on ways of obtaining a better return from Palomar's 18-inch telescope. A report in the planetary science journal *Icarus* reviewed the progress the team had made and its find of 12 new planet-crossing asteroids.[1] How would one get more out of the 18-inch telescope? The solution soon came to Shoemaker's geologist mind: Why not use the same stereographic principle used in some of the stereographic pictures *Surveyor* and the *Apollo* orbiters took of the moon? These pictures gave the features a perception of depth. If he took short exposures and then examined the films in pairs, Shoemaker thought, the moving asteroids would appear not as trails but as single images standing above the starry background. Helin found a California firm called McBain Instruments that built a stereomicroscope capable of holding two 6-inch diameter films at one time. They produced a fine machine for about $10,000— and it was worth every penny. In May and June 1980, Shoemaker and Helin conducted the first tentative observing runs using the stereomicroscope at the 18-inch telescope. The early results were encouraging. By comparing the brightnesses of new discoveries with those of known objects, Shoemaker calculated that they were finding asteroids as faint as the eighteenth magnitude—asteroids perhaps as small as a kilometer in diameter.

Shoemaker and Helin also had a brief flirtation with the 48-inch telescope at that time. Although the first images were badly guided—the stars looked like little seahorses—this telescope later proved capable of finding asteroids even fainter than those studied with the 18-inch telescope. But Shoemaker thought that the smaller telescope was more versatile, since its design allowed faster coverage of the sky each month. "We were coming back from every dark run with lots of new asteroids," he noted.

Feeling more comfortable with the new procedures, Shoemaker added a new twist to the program in fall 1981. In addition to the survey photographs, he and Helin also took films of specific regions in hopes of recovering periodic comets on their way back to the inner solar system. Using both Schmidt telescopes—the 18 inch and the 48 inch—in January 1982, they began a heavy observing schedule.

Because the eruption of the Mexican volcano El Chichon spewed tons of fine particles of sulfuric acid into the upper atmosphere, the sky over Palomar suffered during much of 1982. Helin did discover an Apollo asteroid in February—1982 DB, now named (4660) Nereus. "A benevolent sea-god associated with ancient origins (mythology, if not science)," the citation read, this deity "had the power of prophecy."[2] But if Nereus was prophesying anything that February, it was a change in the program. In fall 1982, Shoemaker and Helin went separate ways, mounting competing programs.

Now working with his wife Carolyn, Shoemaker made several changes. "We stopped taking long exposures, got rid of the filters, dropped down to 4-minute exposures, and really started to cover a lot of sky." The IIa-D film they were using was fast, but a little grainy, resulting in many artifacts and possible new asteroids and comets that did not really exist. By 1983 they switched to Kodak's 4415 technical pan, a slower film requiring longer exposures of 6, 8, or even 10 minutes, but its finer grain was capable of recording fainter objects with fewer film artifacts.

Since then time on the 18-inch telescope has been partitioned equally between the Shoemaker team and the Helin group.

BACK INTO SPACE

Something else brought Shoemaker back into space in the mid-1970s. Lawrence Soderblom, one of his former students and now deputy chief of the *Voyager*-imaging team, paid him a visit. "Gene," he invited, "you paid your dues on the moon. Why not come and join us on the *Voyager*-imaging team?" It was a first-rate opportunity. *Voyager* was the one great hope for studying all the outer planets but Pluto, as well as their moons, in one grand mission. As a multiyear project, *Voyager* was the most ambitious planetary mission ever flown. Two spacecraft would pay calls on Jupiter and Saturn. *Voyager 1* then would visit Saturn's great moon, Titan, before heading out of the solar system, leaving *Voyager 2* to travel outward to Uranus and finally Neptune. At the time of Soderblom's invitation, the spacecrafts were already in the asteroid belt, their cameras calibrated and ready to go. "All the hard work was already done," Shoemaker admitted. "It was an offer I couldn't refuse."

As *Voyager 1* closed in on Jupiter in spring 1979, the imaging team wondered what its coterie of four big moons would be like. Called the Galilean satellites after the man who discovered them in 1610, Io, Europa, Ganymede, and Callisto orbit Jupiter like a solar system in miniature. Earlier observations had already shown that Ganymede, the largest moon, had an icy surface. The big question was, would there still be craters left on it to study, or would the viscous flow of the ice have erased them? Shoemaker hoped that at least one of the Galilean satellites would have a crater record. The bet was that of all the moons, rocky Io would still have its craters.

Not so; Io was a real shocker: That moon had no impact craters—not a single one. In fact as they watched, its surface was being created by eruptions of several huge volcanoes. There were black lakes, calderas, and at least one volcano actually spewing sulfur as the spacecraft tore by. Europa's surface in contrast was glassy smooth with ice. Then came Ganymede, a moon strewn with impact craters and complicated terrain. "Shoemaker!" called John McCauley, one of the team members and a U.S. Geological Survey colleague, "this looks like a pretty good moon for you!" He was right. Ganymede was surprisingly similar to our own moon, with lots of craters, and some of the newer ones had ray systems. Shoemaker took one look at the first Ganymede picture and agreed. This moon had preserved some of its oldest craters. And yet unlike our own moon, all of the younger craters may have been formed by comet impacts. It is possible that asteroids were (and are) very rare this far out in the solar system.

Larger than Mercury and Pluto, Ganymede is the solar system's largest moon. It consists of ancient dark areas crossed by lighter expanses distinguished by long grooves representing its more recent history. Ganymede is a careful recorder through all but its oldest craters of comets that once inhabited the outer solar system. So, it turns out, is the Jovian moon Callisto.

"On Ganymede," Shoemaker explains, "we have an independent source of information of the size distribution of comet nuclei." While both asteroids and comets can make craters on our moon and on other bodies of the inner solar system like Mercury, Shoemaker theorizes that comets dominate the population of small objects in the outer solar system. The only asteroid craters are caused by rogue asteroids that have been put into Jupiter-crossing orbits by repeated encounters with the inner planets. Once an asteroid becomes Jupiter crossing, its fate is almost certainly to be ejected from the solar system.

Craters on Ganymede and Callisto, Shoemaker believes, "tell us something about comets that is hard to find out any other way." The oldest craters may represent an early episode of heavy bombardment by planetesimal objects.[3] One such older feature on Callisto is called Valhalla. Almost 2000 kilometers in diameter, this basin was formed before the surface hardened, its rim resembling a series of frozen waves.

In 1981 the *Voyagers* found a very different story on the moons of Saturn. The inner moons had craters that were generally smaller than those in the Jovian system but with one grand exception: Mimas had one crater, called Herschel, so large that the impact that formed it must have come perilously close to shattering the entire moon. Shoemaker believes only the latest phase of Mimas's history is preserved by Herschel and the smaller craters. Previous large impacts may have repeatedly broken Mimas apart, with gravity reassembling the pieces each time in a different order.

By the time *Voyager* 2 left Neptune and its moons in August 1989, it had unlocked a door to our solar system's past that had not been opened before. To learn about comets, we have to observe comets, but comets in the outer solar system are difficult to follow.

Thanks to the appearance of craters on these moons, we now have a record of what the comet population is like out there.

Voyager's encounter with Neptune took place a few days before one of the Shoemakers' monthly observing runs at Palomar. Since her husband would be tied up at the JPL, Carolyn asked me to observe with her at the 18-inch telescope. It would be my first observing run at Palomar. As I prepared to leave for California, I wanted to see *Voyager's* views of Neptune and its big moon Triton. Tucson's public broadcasting station was carrying *Voyager's* images live, and it seemed like my TV set was connected directly to the spacecraft then speeding by Triton. What a night to remember, I thought. But it was also a clear night. In between views of Triton's crater-scarred surface, I went outside to the backyard to do my thing—that is hunting for new comets. At 9:00 P.M., I opened my backyard observatory and aimed Miranda, my 16-inch telescope, on a patch of sky in the west. For a half-hour, I'd scan, then I would go inside to catch a few minutes of *Voyager* as it sped by Triton. Back out to the observatory, back in again.

It would have been simpler to just spend the evening in front of the tube watching the most riveting broadcast since Armstrong's 1969 walk on the Moon, but I was glad I didn't. Just past 11:00 P.M., I nabbed my fifth comet, Okazaki–Levy–Rudenko, 1989r.

⚉ 13 ☽

Eagle-Eyed Carolyn
30 Comets and Counting

Five hundred miles west of the site from which I found Comet Okazaki–Levy–Rudenko, the sun was about to set over Palomar Mountain in Southern California. At Palomar this is like those few moments in a theater when the audience awaits the curtain rising. Slowly, majestically, the dome shutters at this magnificent observatory move apart, revealing the darkening sky whose brightest stars are already tuning up for the overture. The smallest of the four domes houses Palomar's 18-inch Schmidt, a telescopic camera whose films can record large sweeps of sky—almost 9 degrees across—on a single photograph. Although I have seen hundreds of these observatory nights, I'm still struck by the wonder of the moment the domes open.

Only 2 days after my comet discovery, I experienced this wonder first-hand, for I had gone to Palomar on my first observing run. In December 1988 the Shoemakers and I tried observing together with a recalcitrant Schmidt camera in the Catalina Mountains north of Tucson, but now I was at their Palomar site. I walked into the small observatory building to be greeted by Carolyn's warm smile. "A new comet, David!" she said. "Congratulations!" Since the sky was getting darker, we didn't talk too much about the comet, except that she did say we would photograph it tonight.

But it was time to get to work. Carolyn taught me to load film into the filmholder and then walk up the 12 steps to the dome and load the filmholder into the telescope. "It's a little tricky," she explained. "First you open the telescope doors, then you do this, then you do that, and then a miracle occurs—and the filmholder is ready!" Loading the film into the telescope is something you have to do by practice and feel, not sight. But although I practiced with an empty filmholder before dark—no miracle yet—I was awfully slow when the observing started. The next day I loaded filmholders again and again, and suddenly the miracle occurred—I was doing it right and fast. By the second night, I felt a part of things.

Within a few months, I was getting pretty used to the routine, and Gene, Carolyn, the telescope, and I were becoming a real team. I was getting better at guiding the telescope. I was also learning that photographic and visual observing are equally exciting but in different ways. And the time I notice that the most is when I am actually guiding on a star in the eyepiece.

To prepare for that, Gene loads a film into a special filmholder in the downstairs darkroom and walks upstairs with it; he hands it to me. I install the filmholder in the telescope, lock it in place, set the focus, remove the filmholder's cover, and close the telescope's doors. Gene calls out two numbers, the right ascension (the celestial equivalent of longitude) of the new position and its declination (latitude). I move the telescope and center it on a star. When I am ready he counts down, "five, four, three, two, one, open!" I open the shutter, and the film starts to gather light.

For the next 6, 8, or 10 minutes—the exposure varies according to the brightness of the object we are after—I really appreciate the difference between visual comet hunting and its photographic counterpart. If I were using my own telescope, I would be watching field after field of stars move through the eyepiece, and I might pick up an eleventh-magnitude or brighter comet. But now I concentrate on a single solitary star. In this curious exercise called guiding, I must keep the star exactly centered on the crosshairs of the eyepiece. It is a challenge, for even though the telescope is electrically driven to follow the star, it often lurches forward or

Eugene and Carolyn Shoemaker and the author at the 18-inch Schmidt telescope at Palomar. (Photograph by Terence Dickinson.)

backward a bit. If I can't keep the star centered perfectly on the crosshairs of the guiding eyepiece, the resulting film will show jerks and wiggles instead of starry points of light. This is a battle between me and the telescope, with the distant star as referee. For 8 minutes I freeze the sky on film and only imagine the wonders that may later be found there.

Gene is back with a new film he has loaded in the downstairs darkroom. Ending my reverie, I close the shutter, move the telescope to a horizontal position, cover the film, remove the filmholder, grab Gene's new filmholder, load it, focus the telescope, point the telescope, and find another star to guide on—all in less than 2 minutes. Gene's countdown starts a new exposure, and I acquaint myself with a new star all over again. After each set of eight films has been taken, they are developed.

A perfectionist by nature, Gene loses patience quickly when things do not run smoothly and quickly, so I try to learn fast. The 12.0 steps I climb to the telescope each few minutes don't seem like much, but after 8 or 9 hours of climbing, the decimal point disappears, and I start getting really tired. But despite the pressure of trying to take as many as 70 good photographs on a long winter night, I love the experience. There is a lot of pressure but also a lot of good humor; Gene and Carolyn are a lot of fun to work with.

COMETS BY CAMERA

"One ounce of aluminum and photographic emulsion," Helen Wright wrote about Palomar's biggest telescope, "constitutes the essential optical surfaces of the 200-inch telescope. To permit the surfaces to function properly, however, some 35,000 lbs. of glass and about 1,000,000 lbs. of steel are required."[1]

It is a miracle of optics that the whole purpose of all those tons of metal is to keep a microscopically thin coat of aluminum in the right shape and at the right distance from an equally microscopically thin, light-sensitive emulsion. Although it is a much smaller telescope, the 18-inch Schmidt camera works the same way. Named after its Estonian inventor, Bernhard Voldemar Schmidt, this type of telescope uses both a mirror and a specially matched glass lens to provide photographic views of large areas of the sky—in our case, the field of view is 8.75 degrees (or about 16 full moon diameters) across. It is the resulting film that Carolyn examines in the hope of finding a comet or asteroid.

As we discussed earlier, Edward Emerson Barnard began the photographic tradition by finding Periodic Comet Barnard 3 (1892 V) as a trailed smudge on a photographic plate. In 1981 a space satellite called *Solwind 1* found a comet that crashed into the sun. The several *Solwind* satellites, the Solar Maximum Mission, and the *IRAS* satellite all detected comets, and with the catch of Comet Spacewatch 1990x on a CCD array, that legacy has expanded into the modern field of electronic detectors. The visual finds are made mostly by amateur astronomers, and the photographic detections are made mostly (but not always) by professional astronomers. They do not often compete. Most of the visual finds are in areas relatively close to the sun. The visual searchers also avoid the richest parts of the Milky Way. Photographic discoveries tend to be in areas far from the sun, of comets usually too faint for small telescopes to detect. A single comet rarely bears the names of professional and amateur observers. In 1968 Mark Whitaker, a 16-year-old amateur astronomer from Texas, began comet hunting as a summer project with his 4-inch diameter reflector and found a ninth-magnitude comet after only three evenings! Only two nights later, Norman Thomas used Lowell Observatory's 13-inch telescope with which Pluto had been found in 1930 to discover the same comet, and the comet was named Whitaker–Thomas.

COMET HUNTING AT A DESK

Using a special stereomicroscope, Carolyn Shoemaker scans each pair of films. Just as in any stereophotographic pair, any object that is in a different position from one photograph to another stands out clearly, not unlike those stereo movies people saw during the 1950s. Two hundred years ago Caroline Herschel turned a career of assisting her brother into a comet-finding vocation of her own. Now the Shoemaker effort, which began with her husband's ideas, became a team effort whose findings are in large part due to the eagle eyes of Carolyn Spellmann Shoemaker.

FROM CAROLINE TO CAROLYN

After Gene and Carolyn were married in 1951, for much of the next two decades, Carolyn was a homemaker. "Going through the period of my life when I was raising the family," she recalls, "was just great." Near the end of this stretch however, she became restless. "I would feel like that there had to be something else that was really stimulating for me out there. What I really wanted was something that would absorb all my time and energy like Gene's work does his."

With Gene often away with field geology or work on lunar missions, Carolyn found their Pasadena home getting a little lonely. In the middle 1970s, Carolyn ventured out, but not immediately, to work with her husband. For 2 years she learned a lot about plants, not planets, by working in a flower shop. Although this transitional career was fun, she never intended it to be permanent. In the late 1970s under the guidance of Bobby Bus, an undergraduate student at Caltech, Carolyn began to look at photographic plates that had been taken with a large Schmidt telescope in Australia. Using a blink comparator, she compared images taken on one plate with those on another.

A blink comparator is a device designed to compare plates taken of the same areas of sky. Conceived by Max Wolf in Heidelberg early this century to help in his survey of asteroids, it involves looking through an eyepiece while a small motor alternates the light source from one plate to another. If a moving object is on the plates, the observer can detect it as it appears to move back and forth. There is an apocryphal story that Wolf had also conceived the idea of using the stereomicroscope so that both films could be studied simultaneously, but he abandoned that idea because he had good vision in only one eye. In any event, the blink comparator was the instrument of choice for photographic searches for many years, becoming famous when Clyde Tombaugh used it to discover Pluto in February 1930.

At first Carolyn was uncertain about using the instrument. "I was tremendously nervous the first time I used even a simple calculator," Carolyn remembers. "I thought I would make a mistake and the whole thing would blow up!" But by 1980 she was gaining confidence, and she joined the Palomar planet-crossing asteroid survey, a project directed by her husband and Caltech astronomer Eleanor Helin.

In 1979, Gene developed a plan using a stereomicroscope. Instead of blinking on and off, a moving object now appeared to rise above the background of stars. The process is the same as looking at a pair of identical photographs of a landscape; when viewed in stereo, the nearby trees force themselves dramatically in front of the distant mountains. Developing a proficiency with this new instrument was quite a learning experience. At first both Shoemakers anxiously looked at their pairs of films, discovering every possible speck of dust and emulsion defect. But with some experience, Carolyn quickly learned to differentiate between what was real and what was not. With experience, she says, "your eye and your mind will tell you when you have something." Occasionally that something might be in an area she looked at a few seconds ago. Moving slowly through the field, she would suddenly become alert and move back to check.

"Carolyn went gangbusters on the stereomicroscope," Gene raves. "It was a natural thing for Carolyn, who has a high ability to pick small things out, to recognize small anomalies." In 1982 Carolyn and her husband started working together at the 18-inch telescope. For Carolyn this was an even more basic change. "I am a morning person," she notes, "so I was uncertain whether I could stay up all night and work with the telescope." By early 1983 the Shoemakers were getting used to the 18-inch telescope, and Carolyn was enjoying the stereomicroscope thoroughly. One afternoon while scanning two films they had taken a few nights earlier, Carolyn came across two small hyphens in the sea of surrounding stars. Looking first at one film and then another, she realized that something was moving across those films, and fast. "We were both elated to find something like that." Later named Nefertiti, this was an Amor-type asteroid whose orbit crosses that of Mars. It was the first of a long series of near-Earth objects that Carolyn would discover through the stereomicroscope. Later in 1983 Carolyn found a second Amor, designated 1983 RB and seen again 10 years later.

Carolyn's first view of a comet was a disappointment. Found by Ted Bowell in 1980, it was far away and faint when she first saw it on her films 2 years later. As a result of its encounter with Jupiter, Bowell's comet was off on a new orbit that would take it out of the solar system forever. "It showed up pretty well on one of our films," Carolyn noted, "but not on the other. I thought that if they are that diffuse and faint, I'm never going to discover a comet." By 1983 Carolyn had observed several comets, including a few Comets IRAS, detected not by a human directly but by the Infrared Astronomical Satellite. Learning that not all comets were as difficult to see as Comet Bowell, Carolyn started anticipating the excitement she would feel if she ever found one of her own.

The time for wondering came to an end on a September afternoon in 1983.

> Gene was out of town, and I was trying to get through what seemed like a monumental number of films before we went observing again. And all of a sudden, there it was, and I knew it was a comet. I had a feeling that it might be a new comet, but I wasn't sure. We didn't have a catalogue of comet positions at the time, or anything else to tell us if a comet was known or not. So, I hastily wrote down the approximate positions, plotted them on my star map, telephoned Brian Marsden, and said, "I have a comet."[2]

Comet Shoemaker 1983p was the first of a procession of comets. By 1987 Carolyn had found eight, surpassing Caroline Herschel as the woman who found the most comets. "Passing Herschel's record was a special goal for me—not because there was anything personal at all there, but because it was a landmark and special in a way to find more than any other woman had found so far," she told me. With her fifteenth comet find just 2 years later, Carolyn surpassed William Bradfield's 14 comets, her second goal. From Adelaide, Australia, Bradfield had more comets than any other living person. But other than comparing totals, the methods of visual versus photographic hunting are quite different. "I have just the greatest admiration for the kind of comet hunting that Bradfield does," Carolyn insists.

> Visual comet hunting takes an awful lot of patience that I'm not sure I would have. It involves a lot of cold hours and a lot of discomfort, and a lot of looking before you find anything. What I do involves a different sort of persistence. But the stereomicroscope is a better tool than many amateur comet hunters have.[3]

SURGING TO THE TOP

With the discovery of Periodic Comet Shoemaker–Levy 4 in February 1991, Carolyn surpassed Brooks's long line of 21 comets, and a year and five comets later, she broke Pons's all-time record of 26 named comets. In 1991 our team found a record number of seven new comets. "I am enormously proud of her," Gene brags. "As long as we have been together, I have known that eagle-eyed Carolyn is a great observer."[4]

With so many discoveries, it might be difficult to remember the stories of each. But some, like Comet Shoemaker, 1984o, stand out. Carolyn had been scanning films all morning and then went for lunch with her husband.

> After lunch, I went back and put a pair of films on the ste-reomicroscope. I looked through the eyepiece, just to see if the films were lined up, and there was this glorious comet sitting there. My heart leaped; it was faint, but to me this comet looked absolutely spectacular.[5]

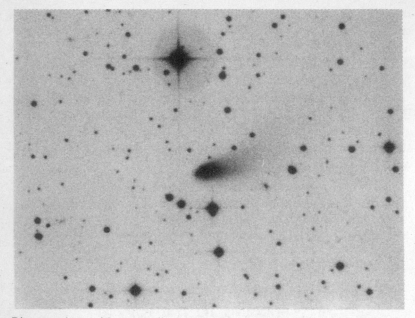

Discovery picture of Comet Mueller (1993p). This comet was discovered by Jean Mueller on a photographic plate taken for the second Palomar Observatory Sky Survey by Mueller and J. D. Mendenhall, using the 48-inch Oschin Schmidt telescope. One of the world's most successful comet discoverers, Mueller has nine comets to her credit.

A month after the Shoemakers found their comet that had split from Comet Levy, Henry Holt, his son Hank, and Tim Rodriguez took a series of films at Palomar, including the field of an asteroid that the Shoemakers hoped to observe. Back in Flagstaff and anxious to know if the films had indeed recorded the asteroid, Carolyn asked Henry if she could examine his films. In no great hurry to scan a film that was dense with stars in the richest part of the Milky Way, Henry agreed. "The films were almost wall-to-wall stars," Carolyn explained. "I was plowing along and hoping to find that asteroid."

When films are so crowded with stars, scanning is difficult, but in less than an hour, Carolyn found the asteroid. She plotted its position and then kept on looking. "Not far away was a comet," she grinned. She called her husband, who confirmed by checking a catalog that this was a new comet. After they telephoned Marsden, Gene smiled broadly at his wife. "Carolyn," he said, "it's June 24. Happy birthday!"

After examining some 8000 pairs of films in the last 13 years, it now takes Carolyn only 20 minutes to scan each pair of films, and she averages about 100 hours of searching per comet. Although this is half the average time reported by most visual searchers, it does not include the time spent exposing the films at the telescope, developing them, or preparing them for scanning. It also doesn't include the 500-mile drive the Shoemakers take just to get to their telescope each month.

After 30 comets Carolyn still experiences that nerve-jangling thrill with each discovery. "I do try to contain it for a while," she says, "at least until we find out if the comet is already known. But when I see a comet, my heart gives a big leap of joy." The day will never come when comet finding is routine; each new comet brings with it the freshness and thrill of that first discovery.

❦ 14 ❧

Are We the Progeny of Comets?

Imagine a large dark cloud sitting passively in space for an incredibly long time. Known as a giant molecular cloud, it is more than 300 light years across—one of the largest objects in the galaxy. It is also one of the coldest, consisting of molecular hydrogen hovering near absolute zero. Somewhere nearby a massive star filled with heavier atoms like carbon suddenly runs out of material for nuclear fusion. In a fraction of a second, the huge sun collapses on itself, and then blows apart in the awesome explosion of a supernova, spreading carbon into nearby space and into our giant molecular cloud. With the injection of carbon and the possible help of some ultraviolet irradiation, the cloud slowly begins to evolve into a different place. Over millions of years, particles heat up and recombine spontaneously into organic substances.

WHEN STARS EXPLODE

It is said that in 1572, Tycho Brahe asked his neighbor to pinch him as he looked at a completely new star in Cassiopeia. Bright as the planet Venus, this star had simply appeared out of nowhere. Viewed from outside the galaxy, Tycho's star would have been as bright as the rest of the Milky Way galaxy. In this age before the telescope, almost 2 full years passed before the star faded from view. In 1604 Tycho's student, Johannes Kepler, saw a second bright new supernova that also remained bright for a long time. After that our galaxy and its closest neighbors, the Magellanic Clouds, remained quiet for almost four centuries. In the long years since then, through the development of the telescope and the spectroscope, which can analyze the light from a star, there was no nearby supernova to study.

At least there was none until February 1987. The best place to be at that moment would have been the office of the CBAT, where Marsden had quietly settled down to work. Just before 9:00 A.M., the telex machine in the corner sprang to life with a message suggesting that a supernova had appeared in the Large Magellanic Cloud, which at a distance of 169,000 light years, was one of the galaxies nearest to ours. Actually the message was somewhat vague, and as he was thinking about making further enquiries, a telephone call from Chile brought complete assurance that astronomers Ian Shelton and Oscar Duhalde had independently found the first bright supernova in 380 years. Seconds later CBAT administrative assistant Donna Thompson took a call from Rob McNaught in Australia with further observations of the fifth-magnitude star. It later turned out that variable-star observer Albert Jones in New Zealand had been an independent discoverer. "It is very important in this business to give proper credit where it is due," says Marsden, "for otherwise the historical record can so easily be distorted."

By 11:00 A.M. CBAT was a madhouse. Telephones were ringing and astronomers from all over Cambridge's Harvard–Smithsonian Center for Astrophysics come by to see what was going on. Then the first spectroscopic observations came as night fell in South Africa. As new observations came in, Marsden had to keep revising his important announcement circular, but when it finally went out quite late in the afternoon, all the significant details were there.

Comet Bradfield, 1974 III. (Photograph by Jack Newton, using a 12.5-inch f/4.3 reflecting telescope, 10-minute exposure with 103AF film.)

A different view of Comet Bradfield, 1974 III. (Photograph by Jack Newton, using a 450-mm f/8 telephoto lens, 5-minute exposure with Tri-X film.)

We learn so much by watching from the sidelines as a nearby supernova blazes out, for it is in the very moment that a star evolves to its limit of nuclear fusion and collapses that all of the heavier elements necessary for life, like oxygen and carbon, are released into space. But although they are the most dramatic source, supernovae are not the only source of these life-starting elements. Large, cool, carbon-rich stars also have contributed these materials more gradually.

THE BIRTH OF THE SOLAR SYSTEM

Let us return to that giant molecular cloud, still made mostly of hydrogen but now enriched with heavier elements from the supernovae and possibly carbon stars. Gradually a fragment of this cloud became unstable and started to collapse, the force of its own gravity causing it to coalesce into a thicker cloud, with its tiny grains covered by a thin layer of organic molecules like frosting on a cake. What caused the cloud to start collapsing? Possibly the shock wave from the supernova itself set it off, or for some other reason, the cloud began to lose its magnetic field and start compressing. It also started to rotate slowly, and gradually the grains settled in the nebula's central plane. The particles clumped together—the scientific word we use is accretion—very slowly as individual particles that settled to the central plane found one another. At the center of the accretion disk was a body that would grow and grow until one day it began nuclear fusion and became the sun. All this took place about 4.5 billion years ago.

Although the accretion disk's organic grains were almost pure carbon, they might have contained some carbon monoxide (CO), water (H_2O), and formaldehyde (H_2CO), which are pivotal materials in the chemistry in which simple organic materials combine to form biological molecules. For example formaldehyde reacted to form carbohydrate sugars. These "prebiotic" reactions could have taken place in the original large molecular cloud, later in the nebula that condensed to form the solar system, or later still in comets.

What happened after that is the subject of considerable debate, and the scenario I present is just one possibility. Perhaps the disk became unstable as more and more grains coalesced into solid clumps of material some 1–10 kilometers in diameter. As the clumps ran into each other, they grew into larger bodies called planetesimals, and into smaller bodies some call protocomets or cometesimals. Some scientists don't believe that scenario. Instead they theorize that planetesimals grew larger until many collisions from smaller objects started to fragment the primordial bodies.

As the accretion disk continued to rotate, its center grew hotter and hotter. As temperatures soared, the delicate organic matter was at risk. Unable to survive the intense heat, it began to disintegrate. Meanwhile the outer part of the accretion disk, which did not get so hot, retained some of its organic materials.

The time scale for the formation of the planets is critical. The entire process between the collapse of the molecular cloud, the building of the planets, and the ignition of the sun probably did not last more than 100 million years. The big gas giants Jupiter and Saturn had to have accreted before the sun turned on its fusion furnace, for soon after that, the remaining gas in the primordial nebula would have been blown away. Uranus and Neptune were probably the last to grow.

Once formed, the infant solar system had a group of warm inner planets without a large amount of organic material; some cold outer planets; the asteroids; and the comets, where organic grains still existed. If life were ever to evolve on Earth, there had to be a way of transporting the still intact organic material at the outer edge of the solar system to the Earth.

THE COMET UNDERGROUND

Were it not for comets, the inner solar system (including the Earth) would still be a place without organic compounds while the outer solar system would continue to house them. The scattering of the comets made the difference. Over a long period of time, they might have provided the underground railroad that brought the organic materials here. Comets provided the Earth with its supply of carbon and its supply of water. Whether they also provided the seeds for life in the form of amino acids is far less certain. In 1982, J. Mayo Greenberg of the University of Leiden predicted that a large fraction of a typical comet would be composed of frozen water—others say as much as three-quarters of a comet is water—and that another large fraction would consist of organic molecules.[1]

While it worked in Greenberg's laboratory, it took an actual visit to determine if these materials were present in a comet. In March 1986 this happened. A flotilla of spacecraft visited Comet Halley, and together with observations from Earth-based telescopes, they confirmed the existence of hydrogen cyanide, water, and formaldehyde. The spacecraft discovered something even more special, that Halley is surrounded by very fine organic particles, now called CHON grains, because they contain carbon, hydrogen, oxygen, and nitrogen. These grains may be the seeds of life, since they contain elements from which amino acids and other compounds essential to life are made. As Gerrit Verschuur explained, CHON is "the simple alphabet of life."[2] (These particles were possibly observed from Earth as jets rich in cyanogen material spiraling out of the comet's nucleus.)

From what we learned from the Halley spacecraft, roughly 11 percent of the comet's mass is carbon. The Earth's crust, by contrast, does not have more than a tenth of 1 percent carbon. But for comets and human beings, the relative amounts of carbon and other materials are close. The University of Toledo's Armand Delsemme points out the closeness from a study of Comet Halley: carbon 9.5 percent in human beings, 11 percent in Halley; hydrogen 63 percent and 55 percent; oxygen 26 and 28 percent; and nitrogen 1 percent and 2 percent. Taken together both comets and human beings are made up largely of the same stuff.[3]

THE COMET EXPRESS

A current line of thinking is that comets were formed in the region moving outward from Uranus. One type of evidence is found in the heavy hydrogen, or deuterium, that they contain. For example the ratio of deuterium to ordinary hydrogen in Halley's comet is significantly higher than what is likely to have occurred in the interstellar medium, the primordial sun, or even the giant planets Jupiter and Saturn. However the amount of deuterium is about the same as that found on Uranus and Neptune as well as Earth. This is evidence that comets did form in the outer solar system beyond Saturn, and it also supports the idea that comets provided much of the Earth's water.

Why would Jupiter not capture the comets instead? Jupiter's gravity was already so strong that it would have flung most of them right out of the solar system. As Uranus and Neptune grew more massive, their gravitational pulls started to scatter the comets much as they both hurled the *Voyager 2* spacecraft out of the solar system in 1986 and 1989. Over hundreds of millions of years, a dispersal of the comets in the region between Uranus and Neptune resulted in an expanding swarm as comets left the hive. Some four-fifths of these comets were ejected from the solar system. Others were ejected directly by Neptune and Uranus into a spherical shell of comets located about a light year away. Named after the Dutch astronomer Jan Oort, this Oort cloud remains as the main storehouse of comets in the solar system. The other half of the original swarm found their way toward Jupiter and Saturn, and some of these went into Earth-crossing orbits. Still other comets were never redistributed at all; these might remain in the outer solar system as a group now called the Kuiper belt, named after the late University of Arizona scientist Gerard Kuiper. It is possible that a few objects recently discovered by David Jewitt and Jane Luu may be Kuiper belt objects. During this period of the dispersal of the Uranus–Neptune swarm, the Earth was bombarded by thousands of these comets, each carrying its supply of organic grains from the outer solar system.

There is one problem with the scenario of the comet express carrying with it the ingredients for life. Comets are also terrible swift swords, capable of destroying the same delicate materials for the origin of life that they help to create. As the comet hits the atmosphere and seconds later the ground, its temperature rises suddenly and rapidly to several thousand degrees, incinerating its supply of organic materials within half a second after impact. It is important to note that the surviving cometary material ends up widely scattered in the atmosphere, where it later settles on Earth.

In March 1989, Cornell University's Paul Thomas suggested a way around the problem of incinerating comets. Perhaps some of a comet's organic materials could survive, he suggested, if the comet were moving slowly enough and if it landed at the bottom of an ocean. During the early period of heavy bombardment some 4.5 billion years ago, Thomas claims, the Earth's atmosphere was 10–20 times denser than it is now—dense enough to slow down a small comet considerably before it hit the atmosphere. The process, known as aerobraking, could slow a 100-meter diameter comet from 25–10 kilometers per second. If the comet then landed in an ocean, it would take less than half a second more to reach the bottom. It is possible that despite the violence of such an impact, the center of the disintegrating comet may not climb above a few hundred degrees Celsius, thus allowing some of the organic molecules to survive. Further if the comet hit the thick atmosphere at a sharp angle, it would be going even slower when it reached the Earth's surface.[4]

A variation of that idea has the comet hitting the Earth at such a slow speed that some of its organic materials incredibly survive an impact on land. This is a very unlikely scenario, but of all the comets that hit Earth, such a "soft landing" would have had to happen only once for this process to spread some organic compounds here. The impact might then form a large crater with a central mound, and it would rapidly fill with melted material from the comet—material consisting of water as well as other organic compounds. Hydrogen cyanide would react to form formamide or formic acid, or nucleic acid bases. As the process continued, the level of this curious lake would fall and rise from seasonal and climatic changes, allowing for periods of drying and rehydration, and resulting in more complicated organic reactions both in the lake and on the wet soil of the central mound.[5] However although this might have happened, a soft landing is not really needed. In a typical crash, the comet disintegrates in the atmosphere, releasing its organic materials to fall gently to Earth, with some of them intact.

We cannot get an idea of the early rates of impacts and cratering directly on Earth, since all the evidence of impacts from this heavy bombardment period has been erased by a planetary lifetime of motion of the crust and rock erosion. However we can get an idea of the rate of early cratering from the moon, whose highland areas preserve the footprints of visiting comets. In astronomical terms, the moon and Earth are so close that they would have been subject to similar bombardments. The Earth could have accreted enough cometary matter—10^{23} grams—to account for all the carbon buried in the Earth's crust and also in all the oceans. The fact that comets are a rather inefficient delivery system, a case of the messenger killing the message, doesn't defeat that argument, since if even only 1 percent of the cometary amino acids or other organic compounds had survived, 10^{20} grams of comet-delivery carbon would still have endured.

Of course there is a big difference between having the ingredients for life and having life itself. Just what brought life to these materials is still a mystery. We are pretty certain that life probably originated after a lot of false starts and accidents. A limerick heard at Oxford sums it up:

> There once was a brainy baboon
> Who always breathed down a bassoon,
> "For"—he said—"It appears
> That in billions of years
> I shall certainly strike on a tune."[6]

Was it a physical electric spark, like lightning, that produced the amino acids that are the ingredients of the primordial soup, or did they come precooked, perhaps by ultraviolet and x radiation from the stars on the CHON grains when they were in space? We have no solid answer for that yet, although there is more evidence now that lightning was not required.

Since Watson and Crick's startling discovery of the structure of the DNA molecule in the 1950s, biologists puzzled over what was considered to be alive first: DNA, RNA, or the proteins. In 1986, T. R. Cech found that RNA itself might be the enzyme catalyst for life, a discovery that earned him the Nobel prize in chemistry 3 years later. The RNA, which stores information like a gene, could have enzyme activity as well. As a living molecule, RNA is probably older than both DNA and proteins.[7]

THE COMET MILK RUN

Comets and asteroids that hit the Earth deposited their dusty grains directly, but many other comets left a huge band of granulated organic matter in a cloud around the sun. We still see a version of that cloud when the evening or morning sky is lit by a peculiar tepee-shaped glow called the zodiacal light. This light covers a large part of the sky, and it is centered along the plane of the ecliptic. This beautiful effect results from the sun's light reflecting off interplanetary dust.

Although current particles of the zodiacal cloud have been orbiting for periods of at least a thousand years, the cloud has had an influx of particles since the beginning of the solar system. In the 1950s, Harvard's Fred Whipple suggested that much of that material might come from comets, in particular, perhaps, from Periodic Comet Encke or a much larger ancestral comet of which P/Encke is but a remnant. However on its way to the outer solar system, the *Pioneer* spacecraft detected a good deal of dust as far out as the asteroid belt, but the dust dropped sharply past the asteroid belt. This indicates that at least a portion of the dust in the zodiacal cloud is asteroidal, not cometary, in origin.

In recent decades we have seen that tiny cometary particles from the zodiacal light cloud, some of which are from ancient comets, actually arrive on Earth. In 1970 the first balloon experiment to collect dust particles in the stratosphere retrieved artificially produced and volcanic dust as well as some microscopic samples of feathery cometary dust. According to Donald Brownlee, who described the balloon findings, these Brownlee particles are extraterrestrial in origin. Specially equipped planes flying some 20 kilometers into the stratosphere routinely collect such particles today. Brownlee particles eventually drop to the Earth's surface at the rate of a single particle on each square meter of the Earth's surface every day.[8]

In the early years of the solar system, zodiacal dust was far more massive than it is now. Over a long period of time, uncountable billions of such grains entered the Earth's atmosphere. They were not large enough to be consumed as the meteors we see in the sky at night. Instead they were so tiny that they came to a stop in the upper atmosphere and then floated down. Drifting gently down in far greater numbers than at present, they added a weighty amount of additional organic materials to the primordial mixture. In fact more of the organic materials of life may have come through this indirect comet milk run, where comets leave their organic residue as individual grains that fall gently to Earth, than by direct impacts.

GETTING THE STUFFING KNOCKED OUT

In addition to providing the substance of life, comets may also be responsible for setting life's house in order. Their carbon, which later oxidized to CO_2 in the atmosphere, began the greenhouse effect that prevented the oceans from freezing. During a period of several hundred million years after the formation of the Earth, a great deal of cometary material was delivered here, either as direct comet hits or as tiny grains left from these comets in the inner solar system. Some 4.5 billion years ago, the onslaught dropped off gradually, then some 3.9 billion years ago, a second blitz swung into action. Just what caused this "late heavy bombardment" is a mystery. Perhaps a single 500-kilometer diameter object, or maybe a large number of objects from the zone between Uranus and Neptune, broke up after a tidal encounter with one of the inner planets. In any event the damage to Mercury, Venus, Earth, the moon, and Mars was immense. From samples collected via the early *Apollo* missions, we can tell that huge impact basins like Mare Imbrium were carved out of the moon by impacts from this period of late heavy bombardment. As Carolyn Shoemaker describes the two assaults: "There was an earlier period when planets were formed, and a later period when the planets got the stuffing knocked out of them."

A WORD ABOUT COMETARY PANSPERMIA

We can take the comets and life idea a bit too far. If comets carried biological building blocks to Earth, could these building blocks actually contain biologically active materials? Could life have originated directly inside comets? Some years ago the British astronomer Sir Fred Hoyle and others suggested that the chances for the formation of even the simplest of enzymes are so low that they could not have formed spontaneously here on Earth. Therefore they proposed, the enzymes must have formed elsewhere in the universe and then been transported here directly by comets. However this part of his argument is self-defeating. Chances of life evolving here are no more likely than for any other spot in the universe, and a very high proportion of the transported enzymes would have been destroyed during the impact with Earth. In short Occam's Razor, the philosophical principle that the simplest explanation should be the correct one, applies, and since having comets transport real enzymes would unnecessarily complicate things, that probably isn't how life on Earth began.

Going further out on a limb, Hoyle expanded his idea to suggest that certain viruses, particularly those that cause common colds and flu epidemics, come from comets. Moreover he suggested that large comets might have caused mass extinctions not from the actual impacts but from viruses that comets spread after impact.

Although current evidence cannot completely exclude the possibility of comets having brought life directly to Earth, the probability of that actually happening is very small. The rhinoviruses that cause colds, for example, are specifically designed to enter the human nose and mouth and then thrive once they get inside. Therefore they would have to have evolved along with their human hosts.

The tapestry of life's beginning is very complicated and leads in any number of directions. We are only now beginning to piece it together. Evidence we have suggests that comets may have played a fundamental role in this picture. As Shoemaker says, "We are the progeny of comets." We now suppose that when the Earth finally formed into a planet, the organic grains were mostly still at the other end of the solar system and over hundreds of millions of years, a large number of long-gone comets carried them past the orbits of the outer planets to the orbit of the Earth. When we gaze at a comet, we may be looking at our own past.

Could a Comet Have Slain the Dinosaurs?

Imagine a peaceful, steamy night with mighty Triceratops *drinking water from a pond, watching for their enemy* Tyrannosaurus rex. Brighter than the brightest stars, a comet dominates the night. For several weeks now, the comet has brightened as it approached the sun, its long tail making it look like a sword in the sky. By the next afternoon, it is all over. With large thunderclaps and a huge crash, the comet slams into the Earth in what is now the Caribbean basin, just off the coast of the present-day Yucatán. Unlike a small meteoroid, which glows brightly as it encounters the atmosphere and is vaporized, the comet is so large that it plows through as if the air weren't even there. High walls of water race out from the point of impact, and millions of tons of dust surge upward in a gigantic cloud. The excavated material rushes out with such force that it quickly circles the Earth. All over the world, the sky is blindingly bright with meteors as the debris reenters; the light and heat make the atmosphere like a furnace. Intense as it is, the meteor storm lasts for half an hour as the surface is bombarded with debris, and surface temperatures are as high as an oven set to broiling. Dry vegetation ignites everywhere (soot in exposed rocks from that time provides evidence of worldwide ground fires). The larger, more slowly moving debris lands a little later in secondary hits close to the main crater, in the gulf of Mexico, causing giant tsunamis to head tens of miles inland on the shores of Mexico and Florida. Soon the whole planet is shrouded in a cloud of dust and soot. The sky is absolutely black, and for over a month, there is no sunlight whatsoever on Earth. The huge amount of nitric oxide results in rain dense with sulfuric acid.

This is an apocalypse, but not the only one: 9 million years earlier, another object struck Earth near the present site of Manson, Iowa.* But the main event was much, much worse. The large dinosaurs disappeared probably within a few days or weeks. Within a year or two, many plant and animal species were gone. Responding to the long-term environmental change brought about by the disaster, some simple but vital plants like algae and nanoplankton disappeared too.

EARLY IDEAS

The scenario just presented is buttressed by the physical evidence in ancient rocks. Had I offered it a generation ago, it would have been ridiculed as fiction, smacks of *catastrophism*, mainstream geologists would have said. I took Geology 100 during those peaceful days of *uniformitarianism*, a geologic school that says the Earth evolved slowly and deliberately, without the need for sudden shocks to its ecosystem, and the dinosaurs died out not over a few months but over millions of years. Between 73–65 million years ago, some geologists still insist, a number of dinosaur species began to decline, due perhaps to gradual changes in the distribution of plants, the climate, or ocean currents. This evidence may concur with new data about an impact. Perhaps the crash was a final

---

* In fact the date of the Manson crater was a subject of hot debate during the writing of this book. As research progressed, various drafts of this chapter suggested the date to be several hundred thousand years later, then earlier, before its current age of 74 million years was determined.

blow to an already declining dinosaur population. Also the idea of asteroid and comet impacts does not mean that the vast majority of Earth's lifeline did not take place in a uniformitarian way.

In any attempt to understand what happened at the end of the Cretaceous period, the sheer magnitude of the carnage is the most important thing to consider. Of the vast numbers of large land-roaming animals that had flourished for more than 150 million years—compared to humanity's 100 thousand year span—not a single one was left. Land and sea-dwelling creatures were affected. Plankton at the ocean surfaces that had been forming the chalky rock at places like the White Cliffs of Dover also vanished.* Some of the smaller reptiles managed to ride out the cataclysm as did some plants. To our great fortune, some small mammals survived also.

The idea of a catastrophe is one of the oldest theories of the dinosaurs' extinction. First offered by Baron Georges Cuvier not long after the first mosasaur remains were discovered around 1770, the idea was that something dramatic interfered with Earth's climate, causing large animals to die out suddenly. The violence of the French Revolution itself, during which scientists were not immune to the guillotine, may have helped put the idea of rapid extinction into Cuvier's mind.

A DINOSAUR DECLINE?

In any case, during much of the twentieth century the idea of a slow extinction reached prominence. There was an argument that dinosaur skeletons were becoming increasingly rare as the Cretaceous/Tertiary boundary approached, but it has been refuted

---

* The Cretaceous period gets its name from these chalky formations that dead bodies of the plankton formed. The chalky rock remained undersea for a long time before finally being lifted above sea level.

by better geologic sampling. At the time that argument supported the theory of an orderly process well.* Paleontologist Robert Bakker, among others, suggests that not enough species died to make the impact theory plausible. Why for example would turtles survive and dinosaurs not? Possibly, other geologists speculate, because they were cold blooded, but the warm-blooded dinosaurs could not adapt to the sudden drop in temperature.

Having survived for so long, dinosaurs are the most successful of the complex animals, and were it not for whatever killed them, they should still be here today. (The recent discovery of *Mononychus*, a predator the size of a turkey but with a keeled breastbone similar to that found in birds, may represent a link between one brand of dinosaur, the *Archaeopteryx*, and modern birds. However evidence for that link is not solid yet.)

One of the earliest theories is that dinosaurs simply evolved themselves out of existence. Some evidence that their skulls became overossified—the development of a strange bone in the back of Triceratops is an example—suggests that extinction may have come about through a kind of species senility. But that idea does not explain why so many species perished at once.

Did chemical warfare kill off the dinosaurs? At the start of the Cretaceous period some 120 million years ago, ferns and conifers provided the major nourishment. The emergence of flowering plants added a new variable however. They produced a strong alkaloid toxic to animals that tried to eat them. By eating too many of these flowering plants, one theory goes, the dinosaurs were poisoned and they perished.

How flowering plants got started is a mystery in itself. Possibly, one theory suggests, the herbivores had grazed on their plants so heavily that these plants disappeared, making a niche for the evolution of new flowering varieties. So the dinosaurs ate off the older

---

* Geologists refer to this geologic boundary as K/T instead of C/T, because C refers to the Cambrian period.

plants and avoided the flowering ones, but with flowering plants showing up everywhere, the dinosaurs adapted to them.

Supporting the idea of a villainous plant is the contorted position that many dinosaur fossils show. With their long necks warped over their backs, the skeletons imply that the animals suffered greatly as they died. Since muscular contraction is a symptom of strychnine poisoning, these contortions do support the plant theory. But there is a big problem with that idea. Why are the remains contorted throughout their history instead of just at the end of the Cretaceous period? Even *Compsognathus*, who thrived during the Jurassic period before the rise of flowering plants, shows these contortions. Actually there is a much simpler answer. The contortions developed after death as the long and delicate ligaments in the neck dried out, slowly pulling it backward.

All available evidence indicates that the emergence of flowering plants did not pose a danger to the dinosaurs. The chemical they released, like sour milk, caused the animals to avoid eating the plants long before the amounts they consumed became deadly. Another point: The dinosaurs reached the height of their development, with their greatest diversity and extent, some 30 million years after these plants appeared. In fact both the dinosaurs and flowering plants coexisted during the entire 50 million years of the Cretaceous period, and mighty *Triceratops* was in full spread at the very end of the Cretaceous.

A SUDDEN EXTINCTION

By 1960, geologists were examining the idea that a drop in temperature, possibly a sudden one, led to the extinction. At this time there was still some question whether the disappearance was sudden or progressive. Catastrophists argued that sampling errors led to the false conclusion that dinosaurs were already in decline.

Toward the end of the 1970s, a new catastrophic theory about the extinction began to make the rounds of geologists. Perhaps the cause was the explosion of a nearby supernova. Chapter 14 discussed the beneficial effects of the explosion of a star, but what if the star was only a hundred light years from Earth and for several months, the supernova shone like a second sun in the sky, bombarding the Earth's atmosphere with enough radiation to knock out most species of life? While typical radiation levels from space average 0.03 roentgen (a unit of radiation exposure) per year, such an explosion would expose the world to some 3000 roentgens inside a few weeks.

This theory sounds plausible except that it is virtually impossible to know if a nearby star did explode precisely that long ago.[1] Victor Hughes of Queen's University discovered two pulsars 456 and 196 light years away that may have affected Earth in a small degree 62,000 years ago and 440,000 years ago.[2] However the remains of an supernova from 65 million years ago would be extremely hard to locate, let alone date. A case of a theory popping up before the evidence was there, this idea caught the attention of the geologic community. But like Einstein's General Theory of Relativity, this is often how science is done: A theory arises, and then scientists look for evidence to support it.

All along our planet was ready to tell us exactly what happened, for the story, like an unread book, was hidden in pages of rock. There was an explosion, but the cause was not a distant star. Two lines of evidence were in these pages, one easily available to some clever geologists who could examine the rocks that marked the boundary between the Cretaceous and Tertiary periods of Earth's history, and the other deeply buried more than a thousand meters beneath the land and water of southern Mexico.

A group led by Walter Alvarez and his father Luis, both from the University of California at Berkeley, found that first line of evidence. In the mountain village of Gubbio, Italy, there is a rocky outcropping of chalk similar in age and appearance to England's White Cliffs of Dover. Atop the chalk is a tiny layer of clay no more than 6 millimeters thick. That thin layer, found in outcroppings around the world, contains more than 30 times more iridium than the surrounding layers. Rare on Earth, the element iridium is more common in comets, asteroids, and meteorites. After a large comet or asteroid hits Earth, the theory goes, a huge amount of dust rises into the atmosphere, then slowly settles down again all over the world. A quarter-inch layer of an extraterrestrial dust rich in iridium—everywhere.[3]

Could the iridium have come from the effects of a nearby supernova, as was earlier thought? The Alvarez paper ruled that out, for among other reasons, a supernova would have deposited some Plutonium-244 with a half-life of some 80 million years, which they did not detect. Also considering the amount of iridium a supernova would emit, we know that the star would have had to be about one-tenth of a light year from us—much closer than the nearest star is today, though not necessarily back then.[4]

To leave a layer of so much iridium all over the planet, a colliding object from space would presumably have had a diameter of about 10 kilometers. Only one known object with that diameter could conceivably hit the Earth sometime in the next several thousand years, and that is Periodic Comet Swift–Tuttle.

THE SMOKING GUN

If a comet or an asteroid that large did hit Earth 65 million years ago, where is the crater? On our plant and water-covered planet, finding it is a daunting task. On the moon, a 100-million-year-old, hundred-kilometer diameter crater stands out like a sore thumb; in fact the rays of Tycho's ejected material virtually go around the moon. The Earth has a very different story. With an atmosphere that erodes geological features constantly and quickly and water that allows new sediment to cover up older features, a 65-million-year-old crater would likely be buried. Given the 75-percent chance that the object hit an ocean, because at least that much of the Earth's surface is covered with water, how could the crater ever be found?

It is part of the beauty of the scientific process that whenever a new idea is presented in a paper, someone else points out its flaws. In the early years after the startling Alvarez paper, Charles

Officer and Charles Drake offered the alternative view that iridium could have come from below, not from above, that is, from the Earth's mantle rather than from an asteroid or comet. These geologists suggest that iridium was deposited by a period of heavy volcanic activity lasting as long as a 100 thousand years. An eruption of the Kilauea volcano in Hawaii, they cite as an example, produced up to 20,000 times the concentrations of iridium found in normal Hawaii basalt.[5] The geologists also suggest that the volcanism could have resulted in cooling and intense acid rain. Recent evidence from India indicates that heavy basaltic flows may have been generated in the 4 million years prior to the end of the dinosaurs.

Arguing against this is evidence that some of the boundary samples are less contaminated with other materials than are others. The purer ones suggest the rapid deposition that an extraterrestrial body would cause, while the more contaminated samples—one even yielded iridium-rich material some 40 centimeters below the K/T boundary—suggest either a slower rate of deposit or deposits from more than one impact.

Noting the lack of an obvious large crater of the right age on the planet, the Officer paper suggested that it would be "encouraging to find some direct evidence of the event itself."[6] So while the impact versus volcanism debate raged on during the 1980s, the search for the crater, and for more evidence from the iridium deposits, began in earnest. By the end of the 1980s, some hundred sites with significant iridium anomalies had been found worldwide. Some of these sites offered a different kind of testimony: They contained small quartz grains that were badly deformed—the result of a shock far too great to have been caused by even a steam explosion from a volcano. This shocked quartz suggested something else: Since ocean basins are composed of basalt, which has a low percentage of quartz, the impact area was on or near a continent and should be still detectable.

By the turn of the decade, Alan Hildebrand and David Kring of the University of Arizona were looking closely at yet another aspect of the K/T-boundary layer. Embedded in the layer at some

Comet Kobayaski–Berger–Milon 1975 IX. (Photograph by Jack Newton, August 4, 1975, using a 12.5-inch f/4.3 reflecting telescope; 15-minute exposure and Tri-X film.)

Comet Kobayaski–Berger–Milon 1975 IX. (Photograph by Jack Newton, August 12, 1975, using a 12.5-inch f/4.3 reflecting telescope; 13-minute exposure and Tri-X film.) Notice subtle differences in the appearance of the tail from the earlier photograph. Particles in the tail are actually moving away from the comet so quickly that changes can appear in as little as 5 minutes.

sites were large amounts of rock fragments that could have been lifted from another place and dropped in their new location by a giant tsunami. Although these giant ocean waves are best known as following earthquakes, an impact of an object from space would have generated tsunamis several hundred meters high, moving hundreds of kilometers inland. Hildebrand noted that these deposits, which were several centimeters thick, cluster around the Caribbean sea and the Gulf of Mexico, an area known as the Caribbean basin.

In 1990, Hildebrand and Kring explored a K/T-boundary exposure in Haiti near a mountain village called Beloc. In its greenish brown clay, they found iridium as well as the same type of shocked quartz grains they had seen in other places. They also found small bits of glass known as tektites—rocks that had melted due to an impact and then cooled rapidly during a brief flight through the atmosphere. Here the boundary layer looked like it was full of material tossed out of the impact spot—so much material in fact that Hildebrand and Kring concluded that the crater could not be very far off.

The missing link was finally supplied by Carlos Byars, a reporter from the Houston *Chronicle* who happened to be at the 1990 Lunar and Planetary Science conference, a meeting held each spring in Houston. Byars had heard of a circular configuration with out-of-the-ordinary magnetic readings 60 kilometers wide, surrounded by a second much larger ring buried under 1100 meters of limestone directly below the city of Mérida and the town of Puerto Chicxulub on the north coast of Mexico's Yucatán Peninsula. Discovered in the late 1970s by Glen Penfield, a geologist from Houston conducting an aeromagnetic survey, the arc was mostly on the water side of Puerto Chicxulub. When he tried to compare the arc with specialized maps made more than a decade earlier, he found gravitational anomalies that showed the land portion of the ring, and he found that the anomalies resembled those of such known impact features as Quebec's Lac Manicouagan.

The story's next stage was a comedy of errors. Although Penfield quickly suspected that the 180-kilometer-wide feature was the buried remains of an impact crater, the Mexican national oil company, Pemex, which had originally ordered the survey, kept the data to itself until the project had been completed. Then Penfield presented his results at a petroleum geology meeting at which few impact specialists were present. Finally Pemex had some material that had been drilled from more than a kilometer below the surface. Before the samples could be confirmed as material that had been metamorphosed by an impact, some were lost in a fire in 1979. When Penfield first examined the specimens, he found that they appeared to be volcanic in origin, but he was puzzled that there was no other indication of volcanic activity anywhere near that region. The melted rocks under Chicxulub were later found to contain higher amounts of iridium than other boundary samples.[7] Chicxulub is an impact crater.

In retrospect, it is strange that the crater's positive identification took so long. Penfield's discovery was written up as a news note in *Sky and Telescope* magazine as early as March 1982, but for the next 8 years, no one paid any attention to it, even though the note included Penfield's suspicion that it might be related to the K/T extinction.[8]

By spring 1991, things were rapidly coming together. Two surviving samples were found to have a large amount of shocked quartz, and images from satellites showed some evidence at the Earth's surface. A large number of cenotes, the local term for large sinkholes that abound there, form a semicircle that matches the ring far below. Perhaps, geologists think, these cenotes resulted as the rim of the ancient crater slumped, producing local collapses in the overlying limestone.[9]

WAS IT A COMET OR AN ASTEROID?

Fifteen years ago the idea of an object hitting the Earth and killing the dinosaurs was so outlandish that few considered it seriously; now the theory is generally accepted. What's left is deciding which of the two types of object could cause that much damage—

an asteroid or a comet. Small meteors appear in the sky every night. Even meteors as bright as the brightest star could be anywhere from sandgrain to golfball sized. A comet nucleus however could be bigger than 10 kilometers across—large enough with a high-speed hit to make a 200-kilometer diameter crater and cause real damage. Asteroids can be that large, too, and during the early 1980s, the prime dinosaur-killing suspect was an asteroid. One reason for thinking this was the amount of iridium in deposits around the world. Since a comet contains a proportion of ices as well as its solid material—basically it has to have room for more types of substances—it would have to be larger than an asteroid to carry the same amount of iridium. Several lines of reasoning now suggest that a comet, and not an asteroid, caused the extinction of more than 70 percent of the species of life on Earth.

Sizes of Asteroids and Comets

Of all the known asteroids presently in orbits that cross the Earth, only one—1627 Ivar—is large enough to make a crater approaching the size of Chicxulub. But even with its 9-kilometer-wide bulk, should Ivar ever hit, the crater it would carve out would be quite a bit smaller than Chicxulub. So if the largest Earth-crossing asteroid we know is too small and the next largest is quite a bit smaller, we have to conclude that statistically the present group of Earth-crossing asteroids is a poor choice for suspects. Chances are that 65 million years ago, while individual asteroids in Earth-crossing orbits were different, their size distribution was about the same as it is now.

Could there be a large asteroid out there with our name on it that has not yet been discovered? Given our present discovery experience with these objects, this is not likely. Asteroid search programs have come across Ivar many times, and if there were another object that large or larger, someone would have found it by now. What if some huge asteroid breaking up somewhere should send a pulse of large, Earth-crossing asteroids toward Earth? It is not likely that such a minor version of the late heavy bombardment previously discussed would have occurred as recently as 65 million years ago.

Comets on the other hand do not drop off so radically as asteroids the larger they get. Both Comets Halley and Swift–Tuttle are large enough to cause the kind of damage that was seen at the K/T boundary, and comets in that size range appear rather often.

Frequency of Comet Appearances

How often do comets in this size range get close to the Earth? Two comets have closed in on us in only the last 200 years— Periodic Comet Lexell, which in June 1770 zipped by at a distance of 1.4 million miles, and IRAS–Araki–Alcock in 1983—virtually just the other day. Comets are always taking potshots at us, and if every 200 years or so, a comet misses us by less than 3 million miles, statistically we may get dangerously close to a large comet, or even hit by one, once in 100 million years.

Noble Metals at the K/T Boundary

Rare elements that do not easily form compounds—metals like iridium, osmium, platinum, and nickel—exist in certain outcroppings at the K/T boundary in proportions that are similar to what we would expect in comets. But since a carbonaceous asteroid (a primitive carbon-rich asteroid) may have a similar "primitive abundance pattern," all this shows is that the impactor was either a comet or a carbonaceous-type asteroid.

A Shower of Comets?

In the western United States, some outcroppings have been found that show more than one iridium layer. Early in the 1980s, some geologists tried to explain this observation as a long series of volcanic eruptions instead of as an object from space. But the existence of several sharply defined layers of iridium seems easily to suggest a series of several distinct events rather than a continuous series of eruptions.

Pockmarking the Earth, many large craters are the signatures of impacts through a stretch of geologic time. Of the craters that have been dated, some are grouped in relatively close time spans of about a million years. A comet shower does not mean that comets fall on us like rain. Over geologic time an increase of 30–100 times the number of hits over a period of a million years would be considered a comet shower.

OTHER MAJOR IMPACTS

In the last 250 million years, there have been four well-documented mass extinctions associated with impacts. Near the end of the Devonian period, some 345 million years ago, a mass extinction, known mostly for the loss of reef-building corals, may be associated with impacts in places as far apart as Belgium and China.

The next major event marked the end of the Triassic period 190 million years ago. Quebec's Manicouagan Crater appears to have been formed at that time and may have been related to that episode. This extinction was actually good for the dinosaurs, for with the drop in population of other animals, the giant reptiles evolved rapidly to fill niches. Tiny mammals also evolved at about this time. After the final extinction of the dinosaurs, this line of life evolved, in a sense, to take their place.

In the mid-Cretaceous period about 92 million years ago, another extinction took place in stages that took about a million years. However the ratio of platinum to iridium in exposed boundary layers is unlike that found in asteroids and comets. It is possible that this extinction was caused by some geochemical episode, such as a long period of intense volcanic activity.

About 25 million years after the great K/T extinction, in the late Eocene period some 40 million years ago, many species of life were extinguished in several steps lasting some 2 million years. The 100-kilometer diameter Popigai crater, east of the Urals in northern Siberia, may be related to part of this extinction. Although the resulting wipe-out took longer, it was almost as gargantuan as the one at the end of the Cretaceous period.

The most recent documented extinction from an extraterrestrial cause took place in the middle Miocene period some 15 million years ago. Limited to a single sample taken from deep waters near Antarctica and another in Australia, the evidence of this is not very strong.

IMPACTS AND EVOLUTION

Could a comet have killed the dinosaurs? Yes. There is a worldwide layer of matter that probably came from space, and a very large crater that appears to date from the same time. Were the dinosaurs already on their way out? Some geologists say yes, and others say no. However whether or not dinosaurs were on the rise, at their peak, or in decline, an ambush from space quite likely ended them.

One last question: How serious are these impacts for the long story of the evolution of life? Darwin's idea of natural selection as a mechanism for evolution portrays an orderly march to more and more complex life forms. But the fossil record adds that the progress of evolution is sometimes reset. While some life forms do evolve to greater degrees of complexity, mass extinctions are often followed by a burst of speciation. Although dinosaurs had been slowly evolving for some millions of years, the extinction event at the end of the Triassic cleared the decks for the dinosaurs to evolve in many directions and probably allowed them to thrive. Had the K/T extinction not taken place, the mammals could not have evolved to fill the niches that were left open in the ecosystem. Whatever the reason for extinctions, they appear to be good for the development of new species—exactly what would follow us however is unpredictable. And if comet and asteroid impacts can be found to have driven most of the major extinctions, then they play an important role in organic evolution.

It is a good thing then that our neighborhood of the solar system is so bombarded. Were there no comets hitting us from time to time, evolution might have stalled out long ago. On the other hand, we are lucky that it is not too active, as it was in the time of the late heavy bombardment discussed in Chapter 14. With K/T-type impacts happening every million years, new species would not have the quiet time they need to proliferate. We can thank Jupiter for that. Throughout much of the solar system's history, this gas giant has been affecting the orbits of many comets, in most cases tossing them right out of the solar system. If Jupiter's enormous gravity were not there, we might have as many as a hundred times the number of impacts, one every million years.

It is also good we don't see too many comets the size of Chiron, a 200-kilometer diameter object that orbits in the outer solar system beyond Saturn. Were something that size to hit just once, it would do so much damage that all life on Earth might be extinguished. Without stopping the process altogether, comets the size of Halley and Swift–Tuttle shake up evolution from time to time, and the dice are thrown again.

☾ *16* ☽

Getting Hit Again

*F*riday, *October 9, 1992 was the kind of clear, warm evening that draws people outside. In the town of Ashland, Ohio, David Hartsel,* a well-known amateur astronomer, was enjoying the Ashland High football game with his children, Heather and Sean.

It was late in the first quarter, and a play had just brought Ashland to the 10-yard line against Wooster. The two teams huddled to work out their next strategies. Suddenly a yellow green fireball emerged out of the twilight high in the southwest. Bright as a first-quarter moon, it raced across more than a sixth of the sky within the next 10 seconds, leaving a bright trail and breaking up into small pieces.

On the field players and referees on both teams pointed to the heavens, and everyone in the west stands saw the meteor fade as it approached the northeast horizon. "Wow!" someone cried out. "What the hell was that? Was it a rocket?" "That was a meteor!" Hartsel exulted. "But it's the brightest one I've ever seen." Play resumed, and almost everyone was home in time to learn on the evening news that people all over the midwestern United States had seen the fireball.

Hundreds of miles to the east, a Peekskill, New York, high school senior named Michelle Knapp thought she heard a car crash right outside her window. When she went to investigate, she found her car's trunk crushed and a warm rock the size of a football next to the gas tank. She reported what she thought was vandalism to the police. But later that evening, geologist William Menke, whom the police had contacted, identified the rock as a stony chondrite meteorite with a high iron content. By the following day, a horde of prospective buyers, including Sotheby's auctioneers from London, were bidding for the meteorite.[1] The event ended happily. Knapp received the grand sum of $69,000 for her rock *and* her car, both of which were displayed several months later at Tucson's international gem and mineral show. And back in Ohio, Ashland beat Wooster 20-zip.

It's funny how nature seems to play deliberate tricks on us. When national networks announced the 1992 Peekskill fireball, they reported that it was a member of the Draconid meteor stream, a logical conclusion, since the Earth was passing through the Draconid swarm that very evening. There had been a swarm of meteors from the same Draconid meteor stream many years earlier, on October 9, 1946. They lit the sky in spite of a bright moon that night. These Draconid meteors were remnants of Periodic Comet Giacobini–Zinner, which, at its return in 1985, became the first comet ever visited by a spacecraft. But although there were Draconid meteors that night in 1992, the great Peekskill fireball was not one of them. It came from the southwest instead of the northeast, where Draconid meteors were centered. Also the fact that the fireball produced a meteorite showed it as the remnant of a tiny asteroid, perhaps less than 1 meter across before it broke up. Although only one piece was found, other debris might have fallen over a 50-mile swath. This means, it is thought, that the meteorite was too dense to have been the residue from a comet.

Knapp was lucky that only her car was hit. On November 30, 1954, Mrs. E. H. Hodges was in her Sylacauga, Alabama home when a meteorite weighing in at 2 kilograms shattered her roof and proceeded to ricochet off her radio set and then bruise her thigh.[2]

Actually the Earth has been bombarded from space repeatedly in the past, and like Damocles, we will be hit again. Damocles was punished for irritating his master Dionysus by having to sit at a banquet with a sword on top of his head. The sword was fastened to the ceiling by a thread. Recently the British astronomer John Davies suggested that if an asteroid were found to be on a collision course with Earth, it should be named Damocles.[3]

Meteorites as large as the ones that bashed Knapp's car or Hodges's thigh are not uncommon visitors. On the afternoon of August 10, 1972, many people in the western United States and Canada saw a bright fireball enter the atmosphere some 60 kilometers above the Earth's surface. Between 4–14 meters in diameter, this object apparently survived its encounter with Earth. Grazing through the upper atmosphere, it went back into space and continued in a new and different orbit around the Sun. It may make another approach to Earth in 1997.[4]

FIND THEM BEFORE THEY FIND US

On January 18, 1991, astronomer David Rabinowitz discovered a tiny object still in space. It was far above the Earth's atmosphere but closer than the Moon. He was using the Spacewatch telescope at Arizona's Kitt Peak in a survey of Earth-approaching asteroids. Named 1991 BA, the object he saw was zipping along at the rate of 24 degrees per day and accelerating fast. By the end of the night, it had speeded up to 60 degrees per day—that's a third of the way across the sky. Estimated to be 10 meters in size, this small asteroid came so close to the Earth that someone wondered if it were artificial. With the Gulf War only 3 days old, perhaps it was some sort of Iraqi missile. But Marsden's orbit calculation showed that it was a small asteroid in orbit about the Sun. "Had it hit," one newspaper reported wildly, "it would have caused extensive devastation." Hardly—however it was some 10 times larger than the Peekskill asteroid and had 1000 times its energy.

Earth may encounter objects that size every year or perhaps several times in a year. Usually they fall unseen over distant oceans or in daylight. While these minor falls are difficult to take seriously, they could have catastrophic side effects. The Peekskill fall was seen by thousands of people, some of whom asked if it were a rocket. What would be the effect of such a sighting on the leaders of a nation at war? Would they mistake it for an incoming missile and then retaliate with a nuclear weapon? What if 1991 BA itself had hit the Earth? Just a day or two into the Gulf War, it could conceivably have changed the course of geopolitical events. An observant amateur astronomer who had seen lots of meteors as well as a slow-moving reentry of an artificial satellite would know by the object's high speed and its color (often yellow–green) that it came from space. Sometimes it might be hard to tell the difference, but military institutions around the world should be able to test the possibility that an explosion in the sky could be from natural phenomena.

HOW GREAT IS THE DANGER FROM IMPACTS?

Besides the consequence of a misidentified fireball, what exactly must we fear from the fall of a comet or an asteroid? That is not an easy question to answer. We have to consider several variables, some of which are well known, others about which we can only speculate. In a seminal paper in 1983, Shoemaker estimated the frequency of impacts of asteroids and comets of various sizes throughout history from the point of view of what size "bang" they would cause.[5] Although he began with the few actual discoveries of Earth-approaching asteroids and comets known at that time, Shoemaker used the size and frequency distribution of craters on the moon to complete his estimate of danger. A crash that is equivalent to an explosion of from 6–20 kilotons of TNT, he judged, would happen as often as once per year. The carbonaceous meteorite that fell near Revelstoke, British Columbia, in the mid-1960s is an example of such a crash. If we can expect this type of encounter once a year, such an object would actually pass through the atmosphere somewhere over North America every 25 years or so.[6]

A 20-meter diameter asteroid probably hits somewhere in the world every 25 years, producing a 200- to 600-kiloton explosion, equivalent to a small nuclear detonation, every 25 years. A 1–3 megaton impact might occur every century and a 10-megaton jolt, resulting from a 100-meter diameter object and duplicating the potential of our strongest nuclear weapons, takes place every millennium. The latest of these impacts occurred quite recently on June 30, 1908, when a small asteroid exploded over the Tunguska River in Siberia. It passed almost directly above the small town of Kirensk but was seen in broad daylight as far as 1000 kilometers away. The fall ignited trees up to 15 kilometers away from the center of the fall and toppled trees for 40 kilometers from ground zero.

All these effects were relatively local, except for two major ones. There were dramatic sunsets all over the world, and the airburst deposited a very large amount of nitric oxide in the upper atmosphere, a substance capable of damaging Earth's ozone layer. Since scientists then did not have today's monitoring devices, we cannot be certain of the extent of the ozone depletion. However observations at California's Mount Wilson Observatory for a completely different project—variations in the sun's luminosity—serendipitously recorded a 30 percent drop in the amount of ozone in the stratosphere between 1909 and 1911.[7]

The Siberian object did its damage without even bothering to hit the ground. As it entered the atmosphere, it encountered such high pressures that at 8.5 kilometers up, it disintegrated in a tremendous explosion. Coincidentally Siberia was hit again less than half a century later, this time by a much smaller 70-ton iron meteorite that left dozens of small craters, the largest about 27 meters across.

The Earth would encounter larger objects at less frequent intervals. A dense, iron-rich asteroid 50 meters wide produced a 30-megaton explosion in 5 seconds when it excavated Meteor Crater in northern Arizona. Something ten times that size, half a kilometer in diameter, would arrive every 30,000 years, producing a 2000-megaton blast.

By some calculations an object 1 kilometer wide would hit every 125,000 years, and a 2-kilometer diameter projectile might disrupt civilization about once per million years on average. Depending on location and time of year, *i.e.*, early spring in the northern hemisphere, where most of the world's crop production takes place, such a strike could set our civilization back, killing millions. An ocean impact, or even a land hit just before winter, so that the following year's food production would not be badly affected, would not be so disastrous. It would also be better if the object didn't break up much before it hit, for if it did, the energy would be spread over a wider area.

In any case a 2-kilometer object would not extinguish our species. How large an object would it take to render *homo sapiens* extinct? Although scientists differ on the threshold size, there is no doubt that a comet or asteroid from 5–10 kilometers across, producing a bang equivalent to tens of millions of megatons, would do so much damage to the Earth that those left alive would begrudge those who had died. Such species killers have arrived in the past about every 30–100 million years.[8]

DEPENDS ON THE SIZE

How many dangerous objects could lie in our planet's path is only part of the equation. What actually happens when one of these things hits the Earth? In January 1993, planetary scientists David Morrison and Clark Chapman classified the different effects as follows.[9]

High-Altitude Disintegration

If the projectile falls apart at least 40 kilometers above the Earth's surface, like the Peekskill meteorite, there is virtually no damage at the surface. The object would be up to 10 meters across, depending on whether it is a small asteroid or a comet.

Blast Damage and Other Local Effects

If the object explodes in the lower atmosphere (as the Tunguska asteroid did in 1908) or if pieces survive to crater the surface, the destruction could be severe over hundreds or thousands of square kilometers. A comet greater than 200 meters in diameter, or a stony asteroid a quarter that size, would penetrate the upper atmosphere and survive long enough to cause some local destruction, including forest fires.

The 1908 Tunguska object managed to do a lot of damage even though it exploded far above the ground and did not leave a crater. Whether an object lands and forms a crater or explodes in the air depends on its density, speed of entry, and angle of entry. If it comes in at a sharp angle, it will travel through much more atmosphere, giving it a greater chance to break up in flight. Some aspects of the physics of comet and asteroid impacts are similar to those of nuclear bombs, except for one important thing: A cosmic impact produces no radioactivity. Since we've witnessed the effects of a nuclear bomb on Trinity Site, Hiroshima, and Nagasaki and we have the results of years of nuclear testing, we can extrapolate what a far larger explosion could do.

Global Effects (Environmental Degradation)

How large does a comet or asteroid have to be before the consequence of its hitting us became global? If a projectile a 100 meters or so wide plummets down, the damage to plants and animals could be great—perhaps life over the area of a large city would be extinguished. A kilometer-wide object would do about as much damage as exploding all the world's nuclear weapons at once. If the object struck a continent, so much dust would enter the stratosphere that it would disrupt life around the world. Either a comet or an asteroid a kilometer long may be about the threshold size to cause a worldwide cataclysm. The exact threshold size depends on the composition of the object, its velocity, and many other variables. I've seen a number of newspaper articles state that "scientists have calculated that an object larger than 0.62 miles in diameter would cause worldwide damage." But scientists haven't calculated the threshold to that degree of precision. The newspaper writers should say about a kilometer or about three-fifths of a mile.

When a kilometer-sized comet or asteroid hits the Earth, the catastrophe could begin with the deaths of thousands—or millions—of people in the immediate area of the impact. They could be killed directly by falling debris, the shock wave, or resulting fires—millions would die from incineration from the tremendous heat that the impact would generate. Over a longer term, the result could be climatic changes over a period of a few months to a few years. The explosion of 10,000 megatons could put enough dust into the stratosphere to cause a significant drop in land temperatures—an impact winter, like the nuclear winter we've been warned of. However as any weather forecaster knows, the atmosphere is affected by so many variables that we have trouble accurately forecasting the weather let alone a sudden climate change. After an impact average temperatures in some places may fall by as much as 10 degrees Celsius. The impact of a kilometer-sized object may also affect the ozone layer significantly.

Mass Extinction (Environmental Catastrophe)

Morrison and Chapman end their scale with the crash of an object between 5–10 kilometers in diameter, from which there would be environmental damage great enough to cause extinction or near-extinction of our own species. The impact winter would be long and severe, with absolutely no daylight at all for months, conditions similar to what happened at the end of the dinosaur era. We'd be gone, and so would most of the other large animals.

HOW OFTEN DO COMETS HIT?

Earth doesn't keep its craters very long. As of 1993 there are only 140 known craters formed by comets or asteroids hitting us at high speed.[10] But Jupiter's large moon Ganymede, where virtually all the impact craters may have come from comets, displays a better record. From it, Shoemaker has calculated that the frequency of hits on any planet in our solar system is approximately inversely proportional to the square of the impacting object's diameter. We should expect to be hit by large comets a lot less often than small ones. Using the Ganymede data as well as observed comets, Shoemaker and his colleague Ruth Wolfe suggest that the Earth should encounter long period comets whose nuclei are a kilometer in diameter or larger only once every million years.

On the same basis, the Earth should meet a 10-kilometer diameter comet, one big enough to produce a K/T-boundary event, every 100 million years. As I mentioned earlier, Shoemaker notes that this prediction is consistent with a K/T-sized comet approaching to within 5 million kilometers of the Earth every 200 years.[11] In the past two centuries, the Earth has been grazed by large comets coming in from distant space. Comets passed closer than 6 million kilometers from Earth several times, including P/Tempel–Tuttle in 1366; 1491 II, 1743 I, and P/Lexell in 1770; P/Biela in 1805; P/Pons–Winnecke in 1927; and IRAS–Araki–Alcock in 1983. Just one month after the 1983 comet, another comet, Sugano–Saigusa–Fujikawa, came almost as close. In the future P/Finlay in 2060 and P/Giacobini–Zinner in 2112 will come within a few million kilometers of Earth, as well as some comets, perhaps, that are still undiscovered.[12]

Comets rounding the sun on near-parabolic orbits that do not bring them back here very often are the most dangerous of the impacting objects, since they can come with almost no warning. Along with these far-ranging invaders, we have to worry about a population of periodic comets, which include Halley, Swift–Tuttle, Levy, and Brorsen–Metcalf, as well as Jupiter-family comets and comets so nearly dead that they are hard to distinguish from asteroids. The chances that Earth may encounter any one of these is about the same as for the nonperiodic comets. In addition there are probably about 2000 asteroids 1 kilometer or larger that are capable of hitting the Earth.

PUTTING THE RISK IN ITS PLACE

If you feel threatened by these predictions, remember that life is full of risks. Even if we stayed in bed all day, the chances of hitting our heads on the night table are very slim, but if we wake up and drive to the grocery store, our chances of dying in a car crash on the way are about 1 in 4 million. At a conference in St. Petersburg in 1991, Clark Chapman offered the statistics that follow. I have added lightning strikes, which are not always fatal, and state lottery wins, which, one hopes, are never lethal. Chapman's figures do not take into account the obvious caveats: Someone who drinks while driving faces a far higher risk of dying in a car wreck than does a safe driver, and frequent flyers have a higher chance of dying in airplane crashes than people who have never flown.

Ways of Dying	Probability
Motor vehicle accident	1 in 100
Homicide	1 in 300
Fire	1 in 800
Firearms accident	1 in 2500
Accidental electrocution	1 in 5000
Comet or asteroid hit	1 in 20,000
Airplane crash	1 in 20,000
Flood	1 in 30,000
Tornado	1 in 60,000
Bites or stings	1 in 100,000
Fireworks accident	1 in 1,000,000
Botulism	1 in 3,000,000

Ways of Dying	Probability
Lightning strike	1 in 4,000,000
Water with EPA limit of TCE	1 in 10,000,000
State lottery win	1 in 15,000,000[13]

The number of deaths in a single accident is another factor we look at when determining how much attention to give the danger. While an airplane accident is very rare, so many people die in each one that public alarm has been high enough to motivate a lot of research on making aircraft safe. There is a far greater risk of dying in an auto accident where the number of people affected per crash is much lower.

A comet impact is potentially the most extreme case of risk versus consequence. Since none of us knows of anyone who has actually died from a comet or asteroid impact, most people are not inclined to take the risk seriously. The possible repercussions of a hit by an object a kilometer or two in diameter are enormous. It would be equivalent to the energy of a full-scale nuclear war and greater than any other natural disaster. Millions of people may die at once, and the Earth as we know it could be quickly transformed into an alien and inhospitable environment. From this statistical point of view, comet and asteroid impacts rank as catastrophes that grip our imaginations.

❰ 17 ❱

Deflecting the Terrible Sword

*A*lmost 50 *years ago, the famous astronomer Bart Bok revealed that the Milky Way was filled with dense concentrations of gas that could* be the birthplaces of new stars. Although these Bok globules, as they are now known, are faint and esoteric objects, they captured the public imagination, thanks to a 1957 science fiction story by astronomer Fred Hoyle, called *The Black Cloud*. As the story opens, astronomers using Palomar's 18-inch Schmidt telescope (the same one used in real life to discover all the Shoemaker comets) spot a big black cloud approaching the Earth. "Such globules are not uncommon in the Milky Way," the astronomer says, "but usually they're tiny things. My God, look at this! It's huge, it must be the best part of two and a half degrees across!"[1] As the drama unfolds, the globule appears to get bigger as it closes in on the Earth. Finally one morning the assistant appears in a great panic: "It's not there, sir, it's not there!" "What isn't there?" "The day, sir! There's no Sun!"

People going outside that awful morning see an eerie sight: It is "pitch black, unrelieved even by starlight, which was unable to penetrate the thick cloud cover. An unreasoning primitive fear seemed to be abroad. The light of the world had gone."[2]

But this is no ordinary globule: In Hoyle's science fiction, this globule has intelligence and the ability to communicate with scientists on Earth, and it is very surprised to find that life forms inhabit an actual planet instead of living freely in space. The globule hangs around for several months before it finally decides to leave, and daylight returns.

We should be so lucky that a comet would be that understanding, were it to decide to envelop the Earth. If we could only reason with it, perhaps it would go away! But Nature doesn't allow that. Instead some people think that we need to come up with more ingenious solutions that would prod the comet or asteroid off its deadly course.

WHAT WE COULD DO

Suppose we found a comet or asteroid ready to hit us in the next quarter-century, a time span close enough to be threatening to a great number of people right now. Here are some sensible things that we could do.

Educate Leaders about Fireballs

The Earth is bound to be hit by something soon. The Peekskill meteorite was harmless except for wrecking a young woman's car and disturbing a football game. However it could just as well have fallen over a country in the midst of turmoil, and it could have easily been misidentified as a missile or even a nuclear airburst.

Fireballs as bright as the one that left the Peekskill meteorite come through the atmosphere somewhere on Earth every week or so. If some nervous government were to respond to a fireball with a retaliatory strike at another country, then the incoming asteroid or comet could trigger a war. As more and more countries acquire nuclear weapons, the chances that one will be launched by mistake increase. A real attempt to educate national leaders and military establishments about sights in the sky could prevent a war started by accident.

Move People Away from the Impact Point

If we should find a large interloper on a collision course with Earth, we would observe it often enough to know its orbit with great accuracy. We might then calculate where it would hit on Earth. Even if we learned that it would land in an ocean, which would be the most likely scenario, we would have to plan the evacuation of shoreline residents to prepare for tsunamis or tidal waves. Evacuation decisions would have to be made and carried out quite quickly, for in all likelihood, we would not know the path that well until a few days or even hours before impact.

Store a Full Year's Supply of Food

Impact winter as a result of a long period of darkness could result in the loss of a full growing season. Food would vanish. If— and that is a big if—we had lead time of a year or more, we could plan to store immense quantities of food. But this would present tremendous political problems. Who would distribute this food on a worldwide basis? How would it reach everyone in every nation? Who would be responsible for making sure it were neither hoarded nor destroyed? Dangerous as it would be, maybe a comet could provide a great incentive and opportunity for the world's nations to pull together to avert a global crisis.

Deflect the Object to Crash in Midocean

If pushing an encroacher to a new orbit so that it would miss the Earth entirely were not possible, perhaps we could use a nuclear blast to nudge it away from hitting a metropolis like Los Angeles and have it land instead in open water. So long as the invader were not too large, an ocean impact would be best. We would then interrupt air and sea traffic and evacuate people on the coastline from the danger of tsunamis. However if a large Swift–Tuttle-sized comet were coming, where it hits would make no difference. It would plow through miles of ocean in seconds. But such an event comes about once every 100 million years. A strike by a 1-kilometer object is far more likely, coming every 100,000 years.

Trying to deflect an object so that it hits one part of the Earth instead of another is an idea with one big problem. We might not know the object's path well enough until it is too late to deflect it. Think of the global crisis that would result if an object aimed at Toronto, Canada, were moved incorrectly and hits Iran instead.

Deflect the Object away from Earth Altogether

Of all the possibilities for dealing with an incoming comet or asteroid, deflecting it away from us is by far the most publicized and desirable option *if* we find the object years, preferably decades, in advance. We could deflect the object with a nuclear bomb exploded near the intruder. A series of such explosions could slowly prod the object into a new path that would avoid the Earth. We would do this slowly, a bit at a time, instead of catapulting the object away all at once. Smaller blasts would be less likely to rip apart the incoming object, thereby making it much more dangerous.

Deflecting the object would likely be the end of a long series of actions that would begin right after someone found something that after several months or more of observations, appeared to be heading directly toward the Earth. Such a proposal is based on the assumption that the possible collision is far enough in the future— several years or even decades—to allow sufficient time to mount a deflecting program. Unless we mount a serious campaign to discover virtually all the kilometer-diameter objects whose orbits cross the path of the Earth, chances are that we will not have the luxury of decades to prepare.

If we find a threatening comet or asteroid, first we would observe it with the best telescopes we had to understand its orbit as well as we could. We could use optical telescopes and with large radio telescopes, we could bounce radar waves off the object to get an even better idea of its orbit and its shape. It would be important to know the shape, partly to make sure that we were dealing with one intruder and not two. At the end of 1992, radar astronomer Steve Ostro took such clear images of the asteroid 4179 Toutatis, then near the Earth, that it appeared to be two asteroids loosely held together by gravity. An asteroid or comet like that would be very difficult to deflect in a controlled way as it entered the atmosphere, breaking into 2 parts and causing even more damage.

By this time—still some years before impact—we should have the impact site plotted to within a 100-kilometer radius—say, somewhere around the Los Angeles basin. But we could do better than that. A small spacecraft would meet the object, take pictures of it, determine its structure and composition, and—most important—leave a beacon to send back signals to Earth. With that signal we could track the intruder with very high precision. We would know its impact target to within a kilometer—now centered, say, in southwest Santa Monica.

Armed with all this information, we are now ready for the dangerous and controversial part. Engineers would send a spacecraft loaded with a 100-kiloton nuclear bomb. The explosive would not blow up on the intruder's surface but about a kilometer away from it, close enough to change its orbit slightly but not so close that it would blow the object apart. To plant a bomb on the surface would be a terrible mistake. If it were an extinct comet, a surface explosion could blow the thing into several pieces, and then each piece would become an active comet! With a bundle of small comets all changing positions as their newly released gases formed jets, we would have lost all control of where they were going.

Instead a bomb placed nearby would gently nudge the mighty object. Moreover the bomb would be set to explode when the intruder was at perihelion, where an explosion would result in the greatest orbit change. After the explosion we would calculate the new orbit to see how effective the first try was. We'd try a second shot a few years later when the trespasser returned to its next perihelion, and after two or three more tries, the object should be in an orbit that would take it safely past Earth. As we can see however, changing the orbit of an object a kilometer across is not a small task.

WHOA! IS DEFLECTION A BAD IDEA?

At this point we need to stop and ask if this whole idea is far too dangerous to proceed with. Even if we knew precisely what effect a nuclear explosion would have on a comet or an asteroid, launching nuclear weapons into space is dangerous business, and exploding them would be a violation of current UN treaties. The idea of a series of nuclear-tipped missiles heading off into orbit is scary. Like the old *Star Trek* episode in which a malfunctioning nuclear warhead plunging back to Earth is disarmed in the last seconds, a program to launch a series of nuclear bombs into space would be just as frightening to many people as the intruding asteroid itself. Moreover we have no current idea of how a nuclear weapon would behave in space near a comet or an asteroid. Would it simply shatter the object, sending a hail of shrapnel down on us that could be even more deadly than if the original object came down intact? The only reason that the idea of deflecting an intruder in this manner is being thought of at all is that there might not be any alternatives.

BEING MORE AGGRESSIVE

To answer some of these objections, some planners are suggesting that we set about learning exactly how a bomb would deflect a comet or an asteroid. We could try nudging a periodic comet each time it passes perihelion just to see how its orbit changes. But even this idea is controversial; in fact some scientists call it stupid. Messing around with orbits of kilometer-wide bodies, they say, could even send one from a harmless orbit into an orbit that could be dangerous to us in years to come.

What about deflecting a comet toward us on purpose? Once we get the technology right, goes one suggestion, we should go to an extinct comet, drill into it to circulate hot water and evaporate just enough of the comet's gases to steer it into a safe orbit about the Earth. In addition to honing the technology of moving comets and asteroids around in space, this opportunity would give us access to the comet's supply of fresh water. But the risks of such an enterprise would be immense. If our aim were wrong, we would bring this supposedly helpful comet crashing right down on Earth. Also should we not worry about preserving the comet's environment too? Perhaps we need a Comet Environmental Protection Agency.

In any case the idea of gradually moving a space intruder would work for an object found way in advance of its hitting us, and one that wasn't much bigger than a kilometer across. If Periodic Comet Swift–Tuttle were about to hit the Earth, we would really be in trouble. Ten times larger, a thousand times more massive, and moving faster than any object we could ideally move, it would be almost impossible to divert. Astronomer Donald Yeomans compared such an attempt to pushing around a tank with a popgun. For this comet we are talking about launching bombs with hundreds of megatons of destructive power on rockets far larger than any now available.

WHAT WE ARE ALREADY DOING

Will a comet or an asteroid hit the Earth in the next 50 years? For asteroids with their 1–3 year orbital periods, there is a way of finding out. But with comets on thousand-year-plus orbits, the assessment is far more difficult. While potentially dangerous asteroids orbit the sun for short periods, allowing us to find them years before an impact, a comet can swoop in on us from the depths of space with no more than a few weeks advance notice—hardly more than the dinosaurs had if they could have seen it coming. If the comet comes from directly behind the sun, we may get virtually no notice at all.

Begging governments to fund comet research because one might squash us is misleading. Let's face it. According to the table in Chapter 16, the chances of anybody we know being hit by a comet are very slim. If someone is going to fund our efforts to find these things, we should offer a different point of view. Those who search for comets and near-Earth asteroids are like the mapmakers of generations past. Wanting to find new routes and new destinations, they made ever-more accurate maps of the Earth. Today our world has expanded into space, and it is interesting and proper to find out what orbits the sun in our neck of the woods. It doesn't cost much to do that. While working toward that goal, we could conceivably find a bullet with our name on it, but that should be a bonus, not an end in itself.

By searching for comets and asteroids that approach the Earth, we will, like the cartographers of past ages, learn a lot about our environs. That is a good thing, and at a few million dollars per year, it doesn't cost too much. The money would go to upgrading the three currently operating projects.

Spacewatch Survey

Find them before they find us! was the call for a program to search for Earth approachers from the summit of Kitt Peak, southwest of Tucson, Arizona. It is the brainchild of Tom Gehrels, a planetary scientist who has devoted his career to studying asteroids and who has discovered five comets as part of various survey projects.

A colorful and lively scientist, Gehrels has what is probably the most elaborate story about confirming a comet suspect. While searching a photographic plate in Tucson, he discovered a fuzzy trailed image that could be a comet. Since the suspect was faint, he thought that the best telescope to confirm the comet would be Palomar's stalwart 48-inch Schmidt, some 400 miles away. He surprised the director of Palomar with a request to travel that very night to Palomar to take a single photographic plate. The director approved Gehrels's request but only for an hour at the start of that very evening. Racing to the Tucson airport, Gehrels caught a mid-afternoon flight to San Diego and rented a car to speed to the observatory. Driving up Palomar Mountain, Gehrels was thinking more about his comet than about his driving. "The road had zigs and zags," he told me, "and I zagged when I should have zigged." The car went off the pavement, leaving its former driver to hitch-hike the rest of the way to the 48-inch Schmidt. He got his picture, but there was no comet on the photographic plate.[3]

In 1982, Jim Scotti joined the Spacewatch effort, and since then he has added enormously to its success. Using the 0.9-meter diameter telescope and a primitive electronic CCD system, Scotti helped prove the system's mettle with 38 recoveries of faint comets since 1982. In one banner year, Scotti picked up a third of the entire year's supply of comets. He also pioneered the development of a means of determining accurate positions of the objects found with the CCD system. Although the project still uses its original telescope, it has a new and much larger CCD system. Under the current system, their discoveries have skyrocketed. Although many of the objects are less than half a kilometer wide, Spacewatch found the first comet discovered with a CCD and discovers as many Earth approachers as everyone else in the world combined.

Although Spacewatch gets more observing time each month than any of the other projects—18 nights centered on the new moon and divided among Gehrels, Scotti, and Robert Jedicke—their system's narrow field means that they cover less sky; however they can see much fainter objects. Spacewatch has found a number of objects with nearly circular orbits about the sun that are so small that they can be seen only when near the Earth.

Planet-Crossing Asteroid Survey

Using the 18-inch Schmidt telescope at Palomar, Eleanor Helin has achieved some spectacular results. Accompanied by a small group of assistants, sometimes including her husband, Ron, during a clear 6-night period Helin will take enough 6-minute exposures to cover a lot of sky. The program has yielded 15 new comets and some 80 near-Earth asteroids, including several asteroids that passed within several million kilometers of Earth—close enough that astronomers with radio telescopes could bounce radar signals off them.

During the Camp David meetings in 1978, Helin discovered an Earth-approaching asteroid that she subsequently named to honor the peace accord between Israel and Egypt. Using the Egyptian sun god Ra, standing for enlightenment and life, and the Hebrew salutation shalom, which means peace, Helin named the asteroid Ra-Shalom in hope that the peace process would continue and an object orbiting in space would forever honor it.

Palomar Asteroid and Comet Survey

Already discussed in previous chapters, the Shoemaker program has yielded 30 comets as well as many asteroids. An interesting byproduct of this search has been discoveries of a large number of Trojan asteroids.

Early in this century, Max Wolf found the first of a group of asteroids so distant that they apparently "shared" Jupiter's orbit, circling the sun some 60 degrees ahead or behind the giant planet along the curve of Jupiter's orbit. Known as Trojan asteroids, they oscillate about the Lagrangian points ahead or behind Jupiter. They are named for characters in the Trojan War. Those on one side of Jupiter are named for the Greek heroes, while those on the other side are named for Trojan heroes. In June 1990, astronomer Henry Holt and I found something completely new—a Martian Trojan, an asteroid some 2 kilometers wide in a Trojan orbit with respect to Mars. Now called 5261 Eureka, the name recognizes the excitement Archimedes felt after making an important discovery. When asked to tell the difference between a crown made of real gold and a fake, Archimedes solved the problem while taking a bath and thinking up his principle that objects of different densities displace different amounts of water. So thrilled was he, the story goes, that he leaped out of his bath and ran out of the house, presumably still unclad, yelling "Eureka! Eureka!" ("I found it!") We hope that other Martian Trojans will have names that express similar joy in making a discovery.

The lucky find of a Martian Trojan excited the community of planetary scientists. There may have been Trojans around Mercury or Venus at one time, but there do not appear to be any now. "The fact that you two found a Mars Trojan is just really remarkable to us," Carolyn Shoemaker remarked. "It is a real unicorn in the astronomical zoo."

ENHANCING THESE PROGRAMS

Each of the three programs has had its own golden moments under the stars. The logical best step would be to build on the expertise that people in these programs have attained over years of observing. Cataloging every object out there that could potentially hit us would be the cheapest insurance against being hit by an incoming projectile. In 1973 we knew of about 12 Earth-crossing asteroids and several Earth-approaching comets; now we see fewer than 200 Earth-approaching asteroids. Since there are an estimated 2000 near-Earth asteroids a kilometer or more across, we have a lot of finding to do. Augmenting the three programs just described would continue that process, and over the next several decades, we would have a far clearer picture of the environment in space in the vicinity of where the Earth crosses in its journey about the sun.

A group of scientists has launched a proposal to speed things up. Project Spaceguard, as it is called, would mount an array of several large telescopes at sites around the world. By searching much of the sky each week, Spaceguard could discover and calculate the orbits of about 95 percent of all 2000 near-Earth asteroids a kilometer across or larger. (The estimated population of earth-crossing asteroids is ten objects that are 5 kilometers or more across, 2000 at 1 kilometer, and at least 300,000 at a 0.1 kilometer.) But the price tag for this program is at least 250 million dollars. Do we really need to know our environment in nearby space that well so soon? Indeed we have the technology to do it. Moreover for a relatively minor investment, we could significantly reduce our chances of having an asteroid or comet come hurtling toward us.

Does the danger justify the expense? We do not always pay attention to risks on a rational basis; airline safety is an example. Even though the annual loss of life in airplane crashes is far less than in auto accidents, each air crash is so visible that we invest heavily in airline safety. Some risks are so high that we do not question the need to prepare for them in some way. We develop ever more complex weapons to defend ourselves in case of war, for instance, in the hope that they will act as a deterrent to war.

SOME PHILOSOPHICAL QUESTIONS

The comet or asteroid impact risk needs to take its place among all these others. Even if we decide that the costs of a search program are reasonable, other considerations persist. Early in 1993, David Morrison, who has chaired a committee to study the problem, posed several important questions[4]:

Is It Appropriate to Interfere with the Natural Process?

After all, if dinosaurs had deflected their comet 65 million years ago, they might still be here and we might still be tiny burrowing animals. Believing that the comet was fulfilling a biblical prediction of Armageddon, some would object to any attempt to meddle with our predestined future. In any event, getting international agreement on what to do could be exceedingly difficult. "What if you made a mistake?" some suspicious nation might demand; "You could deflect the thing on us!"

Should We Protect against Large and Small Impacts?

While the first question is an ethical one, this is a practical one. The Olympian perspective, Morrison notes, is that we could. At one extreme, we could try to protect everyone and everything from a meteorite that hits the trunk of a car, and at the other, we could change the path of a large comet coming to deliver a bang that would destroy our species. Somewhere in between are the 1-kilometer comets and asteroids that would have an effect on the global environment, and the smaller ones that could have severe but local consequences.

Should We Defer Action until A Threatening Object Has Actually Been Found?

Some scientists maintain that we should spend almost no money at all on plotting a defense against an intruder, instead we should concentrate only on searching for them. Then if one were found, we would hope to have enough time to develop a means of deflecting it. This plan might work in the case of an errant asteroid whose discovery would hopefully come many years before it hit us. But it begs the question of the greatest danger of all, which is from a long-period comet arriving with little warning. When Comet IRAS–Araki–Alcock was discovered in 1983, only 1 week passed before it came within a few million miles of Earth. It is hard to imagine any defense we have now that could deflect a speeding comet that close to impact. As we've seen, had the dinosaur comet been a long-period one, their Spaceguard program wouldn't have worked either unless it detected the comet far enough out.

We have to look at the statistics before making a rational decision. One estimate of the bottom line: Each century there is about a 1 in 1000 chance that there will be an impact on Earth of an asteroid 1 km or larger; over 10,000 years the chance increases to 1 in 10. "But if there is a guy out there," Shoemaker says, "we should know about it. It is one of the most predictable catastrophes you can imagine." Unless we search we cannot know whether we are destined to be hit in the next century. With a modest search program, we could at least find out if an asteroid or comet in a short-period orbit were headed our way.

One final question: Should we take the risk seriously?

THE GIGGLE FACTOR

The main problem with the impact hazard is that it seems so terribly remote that people laugh it off. Although there are Chinese records of deaths from probable meteorites, it is a risk from some poorly defined thing out there in space. In their concern over this problem however, some scientists have made alarming statements the press has harped on. Would an object only 500 meters in diameter cause a global catastrophe? Most scientists say no, but there is disagreement. Astronomers using telescopes may insist for example that the greatest danger comes from intruders larger than a kilometer, which happens to be the size of the smallest objects they typically find. On the other hand, people studying ways of deflecting these objects may say that far smaller objects, which are the easiest ones to reroute, pose the greatest hazard.

By the early 1990s, excitement was building for a search-and-destroy program, aimed at first locating all asteroids larger than a kilometer, and then designing great big nuclear weapons to go out and nudge them into more benign orbits. At New Mexico's Los Alamos National Laboratory, astronomers met with nuclear bomb experts, notably Edward Teller, designer of the hydrogen bomb. The result was a severe clash of views and a flurry of bad publicity. One group proposed an armada of ten ground-based missiles, each with a 100-megaton warhead, ready and waiting for the attack of the comet about to hit us with 2 weeks' notice.

The press responded with a battering ram of derision. "Never mind the peace dividend," the *Wall Street Journal* warned, "the killer asteroids are coming!" Quoting one of the meeting participants who shouted "Nukes forever!" the paper depicted the confab's dispute between those who would spend a small amount of money to catalog all asteroids and those who would hunt them down with all means possible. "Killer Asteroid Dooms Earth!" shrieked a San José paper, adding that "if you believe that, Edward Teller and friends have several billion dollars worth of space weaponry to sell you."[5]

With the fuss surrounding predictions that Comet Swift–Tuttle may hit the Earth in 2126, some scientists questioned whether too much was being made of the possible impact. "After all," says comet scientist Paul Weissman, "even if the comet were 15 days late coming to perihelion in 2126, it would miss the Earth, according to one orbit calculation, by about than the distance to the Moon."[6]

"Despite all the hopes pinned on it," a *Nature* editorial intoned, "Swift–Tuttle is resolutely refusing to be the agent of wild destruction. Too bad!"[7] Even the staid *New York Times* mentioned the skeptics' view that the problem "is a result of scheming by astronomers and bomb makers by practicing the kind of threat inflation the Pentagon excelled at in the cold war."[8]

There is no such scheming. Defending Earth against an errant comet is a whole new field of study bringing together scientists who observe the sky, scientists who calculate orbits, and engineers who design weapons and delivery systems. The controversy will rage for some time, long after the snickers have subsided, and we get back to looking at the real promise here. For the first time in the history of all the civilizations ever to live on Earth, we *can* do something about a comet or an asteroid headed for Earth like a dart about to hit a bull's eye.

Over thousands of years, we have come full circle with comets. Feared at first and later studied and enjoyed, they are now feared again as harbingers of doom. But unlike the soothsayers of yesteryear, today, with a considerable amount of planning and a lot of care, we may be able to travel to a comet and change its path so that it sails harmlessly by.

Maxwell's Silver Hammer
A String of Pearls Strikes Jupiter

*L*ittle did I know when I started this book about comets and impacts *in summer 1992 that a string of comets would smash into Jupiter* around publication time. In July 1994 on the twenty-fifth anniversary of the *Apollo* moon landing, pieces of Periodic Comet Shoemaker–Levy 9 will plow into Jupiter, explode in its atmosphere, and release a million megatons of energy.

FIRST SURPRISE: WHAT A COMET!

Its icy interior long since cloaked by an inert crust, in July 1992 a comet wandered so close to Jupiter that the planet's tidal forces tore it apart. As the ancient body traveled away from Jupiter, its pieces moved slowly apart, and a string of new comets was born, their freshly exposed materials brightening rapidly.

Unaware of the event that had taken place so far away, the Shoemakers and I were discouraged about losing another observing run due to bad weather. It was March 1993, and the year had not been a good one for us. Our 7-night observing run in January, we had only one clear night, and the February run had only one clear

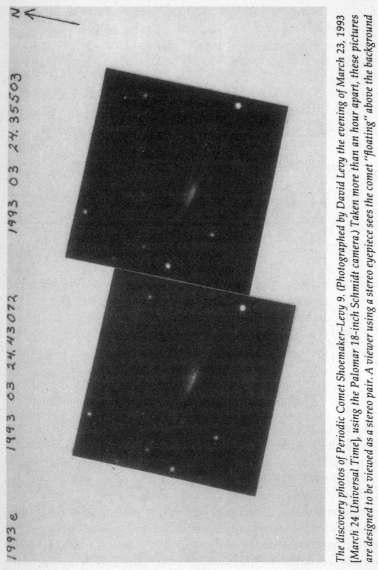

The discovery photos of Periodic Comet Shoemaker–Levy 9. (Photographed by David Levy the evening of March 23, 1993 [March 24 Universal Time], using the Palomar 18-inch Schmidt camera.) Taken more than an hour apart, these pictures are designed to be viewed as a stereo pair. A viewer using a stereo eyepiece sees the comet "floating" above the background stars.

hour out of 7 nights. With poor weather predicted for our March run too, we began to feel despondent. The clouds did give us a chance to work on our other projects. Shoemaker had much work to do for the forthcoming *Clementine* mission that was planned for a 1994 trip to the moon and the asteroid Geographos. Although the little craft would not exactly go excavating for a mine, as the song goes, the observations it made would pave the way for later work that would make *Clementine's* dad proud. And I used the cloudy hours to work on this book.

From the two previous cloudy observing runs at the 18-inch Schmidt telescope at Palomar, we had gathered a collection of blank films, all hypersensitized for 6 hours to increase their sensitivity and be ready to gather starlight in the telescope. Philippe Bendjoya, an astronomer from France who had just completed his thesis on families of asteroids that result when a large asteroid breaks up after colliding with another asteroid, joined us for this week. Finally the night of March 22, the first night of our run, was clear—a breathtakingly beautiful night. The stars were steady, and a marine layer of low clouds beneath our mountaintop hid the skyglow from nearby Los Angeles and San Diego. This was the kind of night astronomers' dreams are made of.

At least it was a dream until I finished exposing the fourth film. By this time Gene had developed the first two films and found to his horror that sometime during the long cloudy interregnum, *someone* had opened the box and exposed the films to light. For a few minutes we thought we had blown the night, for the next films were 6 hours away from being ready for use. Gene quickly developed films from the center and bottom of the box and found that while these were light-struck round the edges, they were still usable. Relieved, we loaded a new film into the telescope and got back to observing.

That turned out to be the only clear night of the entire week. With cirrus clouds coming in at the start of the second night, we hurriedly began our vigil and managed to expose 20 films before the clouds thickened and we had to stop. We stood morosely outside the dome, looking at a sky covered with cirrus clouds. The

Periodic Comet Shoemaker–Levy 9, photographed by James V. Scotti on March 30, 1993 using the University of Arizona's 0.91-meter Spacewatch telescope. The photograph shows about eleven cometary pieces spread out in a bar. A tail extends toward the top from each nucleus. The wings on either side of the bar form a large dust train.

A close up of Periodic Comet Shoemaker–Levy 9, taken by Wieslaw Wisniewski on March 28, 1993 with the University of Arizona's 2.3-meter telescope on Kitt Peak.

clouds were not very thick, I noticed, and I suggested that we continue observing right through them. But with film prices edging $4 per shot, we had to watch the bottom line. Then I had an idea. "Don't we have a dozen light-struck films left? Why don't we just keep going and use them!" Gene looked at the sky again, then at me. "Let's get to it!" he said.

With the telescope horizontal, I loaded the filmholder and focused the telescope—a setting that depends on the air temperature—and Gene read the coordinates as I thrust the telescope toward a seventh-magnitude star right on the ecliptic. "Yuk!" I sneered as a bright glow almost swamped the star. "Dollars to doughnuts, Jupiter's in this field. Okay. Gene, star is centered!"

With a turn of a handle, the telescope's shutter swung open, and for the next 8 minutes, the film gathered light. We took a second film, then a third before the clouds got so thick we had to stop again. For the next hour and a half, we watched and waited. When a small break came, I retook the Jupiter field, and finally we finished the set. Except for a few more fields the following night, we were clouded out for the rest of the week.

By the afternoon of March 25, Carolyn had completed scanning the films from the first night. "I used to be a person who found comets," she chuckled. Shaking her head, she then placed the two films encompassing Jupiter in the stereomicroscope. Suddenly Carolyn sat up tensely. "I don't know what this is," she said. "It looks like . . . like a squashed comet."

This we had to see. Gene stared through the eyepieces, then gave me the most perplexed look I've ever seen on his face. Then it was my turn. Hovering above the stars was the strangest fuzzy thing I have ever seen. It looked like a comet, all right, but instead of the nice round coma that comets should have, this one was a bar of cometary light. What's more, in the hour and 49 minutes between films the whole thing had moved only slightly. Puzzled, I yielded the eyepieces to Philippe Bendjoya.

We puzzled over this for some time. It couldn't be a ghost image of Jupiter, which appeared as a huge black blotch on the film. The ghost image was on the other side of the film where it belonged, and so were some other familiar optical reflections. This was completely different. "The image is most unusual," we reported to Marsden, director of CBAT, "in that it appears as a dense, linear bar very close to 1 arcminute long, oriented roughly east-west. No central condensation [a starry point sometimes marking the comet's center] is observable in either of the two images. A fainter, wispy 'tail' extends north of the bar and to the west." Finally each side of the bar ended with a long, thin line.

Although our sky was cloudy and threatening, I hoped that the incoming storm hadn't moved far enough eastward to cloud Arizona's Kitt Peak, where my friend Jim Scotti was observing that very night. Using the University of Arizona's 36-inch Spacewatch telescope, Scotti was in the middle of his search for asteroids near the Earth. Fortunately the Kitt Peak sky, though "cirrusy," was not clouded out yet. But Scotti worried that the object's motion so closely matched Jupiter's that it could somehow be a reflection of Jupiter in the telescope. While we waited for his attempt at confirmation, Jean Mueller of Palomar helped us measure the comet's precise position.

SECOND SURPRISE: A STRING OF PEARLS

As soon as we returned to the 18-inch dome, I called Scotti. "Have you got a comet!" he exulted. "There are at least five separate condensations!" Later that night he sent this electronic mail message to Marsden:

> It is indeed a unique object, different from any cometary form I have yet witnessed. In general, it has the appearance of a string of nuclear fragments spread out along the orbit with tails extending from the entire nuclear train, as well as what looks like a sheet of debris spread out in the orbit plane in both directions. The southern boundary is very sharp while the northern boundary spreads out away from the debris trails.[1]

The following afternoon Marsden announced Comet Shoe-maker–Levy 1993e. Because the new comet was mimicking Jupiter's motion, we and other astronomers guessed that the comet had been torn asunder in a recent close encounter with Jupiter. One observer however countered (incorrectly, it turns out) that he suspected the original object might have been rotating too quickly and just fallen apart without assistance from Jupiter's gravity.

In the days immediately following this dramatic discovery, Scotti's observations were reduced and fine tuned by David Rabinowitz at Spacewatch, while a host of observers around the world provided early accurate positions that allowed Marsden to compute an orbit. Then Jane Luu and David Jewitt, observing with the 88-inch telescope on Mauna Kea, Hawaii, provided a breathtaking picture that showed 17 nuclear condensations arrayed, they wrote, "like pearls on a string." The University of Arizona's Wieslaw Wisniewski used a 90-inch telescope to take an equally beautiful picture of this unique comet.

THIRD SURPRISE: TWENTY NEW
SATELLITES OF JUPITER

As word spread about the weird nature of this comet, more people started looking at it. Using my 16-inch diameter backyard telescope, I could see it as a slender bar of faint light. On April 3 came surprise number three. Marsden wrote that he suspected that this comet was in orbit about Jupiter, but unlike the satellites moving safely in almost circular orbits about Jupiter, the comet's orbit would take it very close to Jupiter at what we call the perijove, and a year later, as far away from Jupiter as Mercury is from the Sun at apojove, the farthest distance from Jupiter. Other comets have been presumed to orbit Jupiter briefly, but until now no one had ever seen a comet orbit Jupiter this way. Meanwhile Luu and Jewitt had counted 20 comets in the train. So temporarily at least, instead of its normal retinue of 16 known moons, Jupiter had 36!

Marsden later renamed it Periodic Comet Shoemaker–Levy 9, the ninth periodic comet found by our team.

WHAT CAUSED THE BREAKUP?

In the months after discovery, there was some doubt that the comet came close enough to Jupiter to be torn apart by tidal forces. Comets do not necessarily need to be near giant planets or the sun to break up. Comets Levy 1988e and Shoemaker–Holt 1988g, discussed earlier, broke up 10 thousand years ago when they were not much closer to the sun than the Earth is. As Scotti whimsically notes, you can sneeze on a comet, and it will break up. At one point he suggested that a sudden explosion of the comet's volatile material might have helped the destruction along. As it turned out, these ideas faded away once we knew that the comet passed close enough to Jupiter so that tidal forces tore it apart in July 1992.

All this speculation changed by May 22, 1993, however. Using more than 200 precise positions, Marsden was able to compute an accurate orbit that tells of the comet's extraordinary adventure. In July 1992 it passed within 50,000 kilometers of Jupiter's cloud tops, which is well within Jupiter's Roche limit. During the nineteenth century, French mathematician Edouard Roche determined the distance from a planet at which a liquid sphere—or any object without cohesion to hold it together—would be torn apart by tidal forces. An object that passed within Saturn's Roche limit sometime in the last few billion years was completely torn apart and formed the planet's majestic ring system. But until Periodic Comet Shoemaker–Levy 9, we have never seen this effect actually happen. "Never before," says astronomer Clark Chapman, "has there been such a clear case of disruption of a body by tidal splitting."[2] On that day Marsden announced that in July 1994, the comet would collide with Jupiter.

HAS THIS HAPPENED BEFORE?

While Periodic Comet Shoemaker–Levy 9 is a dramatic example of a comet that probably has been torn apart by tidal disruption, it is not the only one. Some famous sun-grazing comets, like the ones in 1882 and 1965, may be traced back to an earlier large comet, possibly the comet of 1106, that broke apart into smaller fragments as it did its hairpin turn round the sun.

In July 1889, William Brooks discovered a comet brightening slowly as it approached the sun. As soon as the comet's discovery was announced, other observers began watching it. On August 1, Edward Emerson Barnard was stunned to see two small companions. He thought that each mate looked like a flawless replica in miniature of the main comet. The next morning Barnard found four or five new objects. Although these single-day wonders vanished by the morning of August 3, two others appeared on the fourth, one lasting a few hours, the other a week. One of the original small comets lasted for several months. Although the main comet has returned 12 times since, the companions were never seen again.

Using logarithmic tables and orbit formulae, orbit computers were able to trace Periodic Comet Brooks 2 back to 1886, when it passed within a distance equal to Jupiter's own diameter from the giant planet. It is the only other comet known to have been torn apart by Jupiter's strong tidal force, and we can compare it to Periodic Comet Shoemaker–Levy 9. But there are major differences: The main body of P/Brooks 2 survived; the other fragments, which were far smaller, apparently did not for very long. Periodic Comet Shoemaker–Levy 9 was catastrophically disrupted after coming too close to Jupiter, forming a substantial number of pieces that must have been bright for half a year or more. Moreover P/Brooks 2 was seen as it approached perihelion near two astronomical units (basically the average distance between the Earth and the sun), while our new comet is putting on its show way out at the orbit of Jupiter.

Some 3.9 billion years ago, the moon and all the solar system's inner planets were hit by a platoon of objects that may have been the result of a catastrophic breakup of a large object. Astronomer George Wetherill has suggested that a body about 500 kilometers across could have passed within the Roche zone of a planet, perhaps Earth or Venus, and broke apart to form a pesky swarm of debris that orbited in the inner solar system, wreaking havoc on all the inner planets, including the Earth and moon. With this new comet, we may be witnessing a small example of that process.

WHY DO THE PIECES APPEAR TO BE STRUNG OUT IN A LINE?

As P/Shoemaker–Levy 9 broke up, Scotti suggests, the large pieces moved out in slightly different orbits. The orientation, he notes, is a direct result of the slightly different orbits that the pieces have. Fragments farther from Jupiter than the comet's center of mass move in orbits with very slightly longer periods than those closer to Jupiter than the center of mass. The slight difference causes the fragments to appear to be strung out in a line.[3]

According to Zdenek Sekanina of the JPL, since the breakup occurred in July 1992, by the time of discovery in March 1993, the pieces were moving away from each other at the rather leisurely rate of 3 meters per second, about a runner's pace.

CAT-AND-MOUSE WITH JUPITER

This remarkable comet probably got into its scorpion's embrace with Jupiter very gradually. It may have begun its journey out in the Oort cloud of comets far beyond the planets or perhaps closer to the sun as part of the Kuiper belt of comets closer to Neptune and Pluto. Over a very long period of time, the comet left its ancestral home and began to travel toward the sun. For many hundreds of thousands of years, the comet roamed in a wide, almost parabolic orbit about the sun. During one of these orbits, it happened to wander close enough to Jupiter so that the giant planet's gravity changed the comet's orbit, and it became a periodic comet. Several approaches to Jupiter later, perhaps as recently as 1971, the comet's orbit now so closely matched Jupiter's that it became locked in the planet's space, looping around the gas giant in a highly elliptical orbit. The comet had become a satellite of Jupiter.

Actually the Periodic Comet Shoemaker–Levy 9 comet club may not, even in 1993, be the only broken-up comet orbiting Jupiter as a group of satellites. Jupiter's moons VIII Pasiphae, IX Sinope, XI Carme, and XII Ananke could be orbiting indefinitely as fragments of a large comet that broke up near Jupiter long ago. (They could also merely be trapped asteroids.) But unlike the unfortunate fragments of Periodic Comet Shoemaker–Levy 9, these four bodies are in stable orbits that could last indefinitely.

As the train of comets approaches Jupiter in 1994, comet pieces will rapidly increase their distance from each other. The train will rapidly become longer, and large fragments will hit Jupiter, although some of the outer dust particles may miss.

FOURTH SURPRISE: COLLISION WITH JUPITER

In these pages we have wondered what damage a major comet would inflict were it to hit Earth. A few days before the May 22 circulars appeared, Marsden sent me a note saying that the orbit he had worked out predicted a collision. I thought at first he was joking. A comet about a kilometer in size hitting the Earth is a once-in-100-thousand-year event. With big Jupiter the frequency is much greater, perhaps once in several decades or once a century. When the circulars actually appeared, I could hardly believe my eyes. Over the next few weeks, a lot of people thought about the consequences. Dismissed at first as "throwing a pebble into the ocean," by CBAT Associate Director Dan Green, the ante was raised as astronomers calculated the energy that the crash would release.

"More like dropping an olive into a martini!" quipped JPL astronomer Steve Edberg. Both are extremes. As the kilometer-wide comets fall toward Jupiter at about 60 kilometers per second, some astronomers speculate, each one will disintegrate in a sudden explosion, releasing as much energy as a million megatons of dynamite. Hiroshima's 20,000-ton blast and the 60-megaton strength of the largest nuclear bomb ever exploded would be firecrackers compared to this awesome cataclysm.

Shortly after the May 22 announcement that a crash was inevitable, A. Carusi of Rome University, suggested that surviving nuclei would continue as satellites of Jupiter or be thrown closer to the sun as independently orbiting short-period comets.[4] But later modeling suggested a different story however—that the whole train, comet after comet, will fall into Jupiter.

The comet's collision with Jupiter is predicted for July 1994, a few months after this book first appears. The material here reflects the state of our knowledge about the comet and its orbit about 9 months before the collision. But despite the fact that our understanding of this comet and its orbit is changing rapidly, this chapter won't be dated; instead it will offer a historical perspective of the comet during the critical period several months after its discovery but before its impact with Jupiter.

As astronomers anticipated the coming explosions, the story began to shift from the effects on the comet to the comet's effect on Jupiter's atmosphere. Already very turbulent with Earth-sized storms and the much larger Great Red Spot, the Jovian atmosphere is divided into belts and zones containing many ever-changing patterns. What will the sudden addition of material and energy from 20 exploding comets do to this churning mass? Astronomers and physicists in the midst of figuring out what a large comet would do to Earth, will now have a natural laboratory and find out. For Periodic Comet Shoemaker–Levy 9 may be compared to some of the great impacts in Earth's past. For the first time in our history, we have the chance to watch what happens when a comet strikes a planet.

EPILOGUE: SHOEMAKER-LEVY 9 COLLIDES WITH JUPITER

Sixteen months passed between our discovery of Comet Shoemaker-Levy 9 and its violent end in the clouds of Jupiter. In January 1994, more than 200 scientists met in Baltimore, Maryland. Because the event of a collision was so new they had little idea of what to expect and planned many different types of observations. During the remaining six months, both observers and theoreticians continued to prepare for this event.

July 16, 1994 came quickly enough, and the world was watching. Virtually every telescope on Earth was armed with detectors set up to observe the impacts, including the Hubble Space Telescope in orbit around the Earth, and the Galileo space probe on its way to Jupiter. Little did anyone know that this was an event that would reward almost every observer.

Just after 4 P.M. Eastern Time, Fragment A approached Jupiter's southern hemisphere at 60 kilometers per second or 135,000 miles per hour. Brightening rapidly as it hurtled through the stratosphere, it began to shatter, then somewhere below the clouds it completely blew up. A fireball shot upward to the incredible height of 3000 kilometers above the clouds, and lasted for almost a quarter of an hour. It climbed so high that Earthbound telescopes were able to see it, even though the fragment hit beyond Jupiter's sunrise horizon. Finally, when the planet rotated enough so that we could see what happened, it displayed its damage—a large dark spot half the size of the Earth, and growing.

The big reflector at Calar Alto, in Spain, was the first telescope to report the electrifying news, and within a few hours they had released their beautiful pictures of the event. Thanks to a message center set up at the University of Maryland, the details of all the early observations were broadcast around the world to astronomers who had been waiting for more than a year. When the Hubble Space Telescope revealed incredible images of the plume from fragment A, as well as a large spot, the observing team cheered and brought out champagne.

Within a day, fragments C and E had also left their marks,

and the world was starting to pay attention. Dramatic images found their way onto the television news and newspapers not long after astronomers had seen them live. But no one was prepared for the cataclysm of Fragment G. On July 18, this large fragment struck Jupiter with such tremendous force that its explosive plume, or fireball, was brighter than all of Jupiter. Fifteen minutes later, a large dark spot rotated on to Jupiter's day side. Surrounding its very dark center was a black ring, and spreading off to the southwest lay a large semicircle of dark material. The complex was as big as Earth. Four other fragments observed by the Hubble space telescope also left dark rings around their impact sites. Thought at first to be expending sound waves, these circles were later interpreted as waves set off deep within Jupiter's water clouds, and expanding outwards at high speed.

The Hubble Space Telescope recorded the spectrum of fragment G as well. One of the most colorful planets in the solar system, Jupiter owes its brilliant reds, greens, and blues to the presence of sulfur at different levels of its atmosphere. Unexpectedly the spectrum recorded the presence of sulfur, suggesting the colliding comet fragment dredged this material up from below.

Finally, early on Friday, July 22, the distant Galileo spacecraft turned its telescope and recorded the last fragment—a part of the comet called "W"—as it tore through Jupiter's atmosphere like an incredibly bright meteor. It brightened, then faded. That was it. The comet was dead.

Who would have thought that young children could look through the smallest telescopes and see the results of the impact of Shoemaker-Levy 9? Before the impacts, we took considerable pains to caution the public that Jupiter would not flare up, nor would there be great changes to see or dramatic events to watch through a small telescope. We were wrong. By the end of the week Jupiter's southern hemisphere was girdled by a series of spots so dark and large that they were the most prominent features ever seen on the planet. It appeared that those spots were made from a soot-like material either coming from the comet or formed by the temperatures and pressures of impact, and they lasted longer than any spot previously observed.

Shoemaker-Levy 9 was a triumph of coincidence. The comet did us the courtesy of breaking apart two full years before impact. This break-up caused it to brighten sufficiently so that it could be discovered and observed in plenty of time for astronomers and observatories to plan observing time calmly and carefully. As of this writing we do not know if every major collision on Jupiter would produce all the dramatic effects that this one did, but we were extremely fortunate to be able to see so much. Never before have we witnessed a major collision in the solar system, the same kind of collision that—were it to happen here on Earth—could be the cause of a mass extinction.

David H. Levy
February 1995

ℂ Epilogue 🌙

Cole of Spyglass Mountain

*M*y love of comets began with a 3-minute speech I gave in sixth grade. I think I said that comets have heads and tails, that they fly about the sun, and maybe even that some people could discover comets. Looking far into the future, I would have said that Halley's comet, the most famous comet of all, wasn't due back for more than a quarter-century.

Within a few months, a fortuitous fall off a bicycle and a get-well present of a book about the solar system got me firmly single-tracked toward the sky. I remember wondering why my little telescope made everything look like a doughnut; then I realized that if I moved the eyepiece, the telescope would focus. Soon the stars were points of light, and Saturn became a ball with a ring around it. Glorious! I had seen the rings in pictures, but to see them for myself was thrilling. I was so excited that I called my parents, who came out to look. It took only a few minutes to see this planet with the rings, but I've never forgotten that night.

For Dad, watching me discover Saturn's ring brought him back to his childhood. One evening over dinner, he told me about a novel he had read when he was a youngster. Called *Cole of Spyglass Mountain*, it was Arthur Preston Hankins' story about a boy whose love of the sky led him to observe Mars through his small home-made telescope.[1] The novel ended with Cole finding evidence of life on Mars one night and becoming an instant celebrity. "If you ever find that book," Dad said, "I'd like to read it again."

Five years later I began my long search for comets. As the years went by and Dad grew older, he would often ask if I had found *Cole of Spyglass Mountain*. Then with the onset of Alzheimer's disease, his memory began to fail. As his time grew short, we spent more of it together, and I watched helplessly as he forgot names, forgot people, and once even forgot that I was his son. But somehow he never forgot that euphoric story about a boy and a mountaintop that had captured his imagination so many years earlier.

When I finally found my first comet in 1984, I felt as though I had rewritten *Cole of Spyglass Mountain* in a private way just for my father. But Dad was too ill to appreciate it, and he died only a few months later. I will always regret that I could not share the most significant moment in my life with the most important person in it.

A few years later, my friend Peter Jedicke found a copy of *Cole of Spyglass Mountain* in a library, and I was able to see its words in print for the first time. Dad had remembered them very well, I found. Though spyglass mountain is fiction, it evokes the magic of discovering new things in the sky. Whether it's a new comet on its way to a meeting with the sun or the remains of a long-departed one as a crater on the moon, the enchantment reverberates through time as each generation first looks toward the sky and wonders about it. Spyglass mountain didn't turn Dad into an astronomer, but it helped him understand what discovery is like, and I like to think that it helped him understand my love of comets. Since then 19 of them have come my way, and somewhere in the excitement of each one, I remember Dad, his long-gone book, and our first look at the sky together.

❆ *Afterword* ❆

How to Discover a Comet

If you are serious about comet hunting, you should first learn to recognize the different types of fuzzy deep-sky objects like galaxies, clusters, and nebulae, both to avoid mistaking them for comets and because they are interesting in themselves. You will find these in the Messier catalog. Representing every type of deep-sky object, this catalog offers excellent training for a would-be comet hunter. Finding these objects at different times of the year and under varying observing conditions will build the discipline that is needed in comet hunting—a skill that does not come overnight but builds on a special love of the sky.

This is more of an invitation to study the Messiers than a warning to stay away from hunting. The Messiers are a *cordon bleu* of the finest deep-sky objects you can behold—the exquisite wisps of M42, Orion Nebula; the nearest spiral galaxy, M31; the spectacular globular M13 in Hercules; the Ring Nebula M57; even the mysterious misshapen galaxy M82—this is the grandeur of the Messiers. If you don't find a good proportion of them before you begin a serious comet search, then you are denying yourself the appropriate introduction to what your nocturnal hunts could offer.

If your study of Messiers gives you an indication of what comets are not, observing comets that are already present and accounted for will give you an idea of what they are. Using positions and descriptions from such sources as the *Circulars* from the International Astronomical Union's (IAU's) CBAT—the IAU's clearinghouse for new astronomical discoveries—or from such magazines as *Sky and Telescope* or *Astronomy*, keep an eye on the comets you can see with your instrument. You will probably find that most comets, even the ones listed in magazines, are quite faint and difficult to observe from light-laden city skies.

Patterns of Search

Although comet hunting can be done with almost any telescope, I recommend one that offers a wide-field view of the sky. Comet hunting works best if you follow a definite plan. Moving the telescope slowly, search a small area of the sky thoroughly, for example, rather than a large one haphazardly. You can also expect better results if you search for 1 hour for 10 nights through areas more likely to have comets rather than searching 10 hours on 1 night through sections of the sky far from the sun and unlikely to produce a comet. Since comets shine brightest when they are near the sun, you have a much better chance of finding a new one in the western sky after dusk or in the eastern sky before dawn. Most important, if you really want to find a comet, don't give up on your quest. During the long years before I spotted my first new comet, I took heart from the advice of Robert Burnham, who discovered several comets during the 1950s and 1960s; he told me that "if you hunt for comets long enough, sooner or later you will find one."[1]

Equipment Required

I have used three telescopes for visual comet hunting, an 8-inch, a 6-inch, and a 16-inch reflector. Your telescope needs to be capable of at least a three-quarter degree diameter field of view. Exactly how large the telescope should be depends on a number of factors: how much money you wish to spend; whether you plan to mount the telescope at a dark site or to carry it around from place to place. The size of your telescope also determines the brightness range of comets you are likely to find. For comets of the eighth magnitude or brighter, a 6-inch telescope will suffice, but such a telescope cannot catch anything much fainter, since the diffuse quality of some comets makes them more difficult to see than clusters and many nebulae of the same brightness.

Methods of Hunting

There are as many variations in comet-hunting procedures as there are successful comet hunters. My telescopes are all mounted on simple altazimuth mounts with motions across and up and down. Canada's Rolf Meier hunts with an equatorially mounted telescope, checking suspicious objects quickly from their right ascension and declination. He never overlaps fields, and he searches at the rate of about a degree a second. In Australia, William Bradfield makes horizontal sweeps with an altazimuth-mounted telescope over a wide arc of sky and always in the same direction. On one point most hunters agree: Patience, skill and luck are necessary ingredients for a successful comet search.

Depending on the peculiarities of your own telescope and its mount, you might try different ways of hunting. You can hunt up and down strips of sky in a zigzag pattern, although this procedure may leave gaps in your search pattern as the setting sky goes past your telescope. You can sweep horizontally along a strip that is at some altitude above the horizon, or you can hunt along specific strips of declination or right ascension. The procedure you follow is not nearly so important as making certain that your search area is thoroughly covered. With my 40-centimeter diameter reflector, I hunt in an up-and-down pattern partly because my telescope moves more easily that way than in azimuth as it crosses the sky.

When to Hunt for Comets

Comet hunting requires some judgment in choosing the best times for sweeping. On successive nights after a full moon, you will see more and more hours of dark sky. Since the areas closest to the sun have the best chance of containing an undiscovered comet, begin searching in the western sky in the nights following a full moon. Until moonrise you have a chance of looking through a sky that has been blocked by moonlight long enough to have masked from other observers the approach of a comet that is getting brighter as it nears the sun or Earth. Although any dark, moonless sky is good, the best time to hunt in the evening sky is just after a waning moon in an hour of darkness after evening twilight ends. Such an event that takes place some 2 days after a full moon. It is important to start your hunt when the western sky is dark enough to show at least fourth-magnitude stars to the naked eye. There may still be light in the West, but your sky is getting darker and so is your adaptation to the dark. Stop hunting when the moon rises in the eastern sky.

Fruitful morning searches begin as the waning lunar crescent is thin enough not to interfere with the dark sky. The adage, "it is darkest just before dawn," makes sense, since pollution from the previous day has had a chance to dissipate, some of the lights from human activity have been extinguished, and the environment is much quieter than on the previous evening. For me the most enjoyable time for comet sweeping is undoubtedly in the morning sky in the last 2 hours before dawn. I begin a few days before and continue for about 5 days after a new moon.

Sun's Vicinity

Would you like to find a comet in broad daylight as it whips round the sun? The procedure is simple and requires only a sharp pair of eyes and a clear sky free of haze or cirrus clouds. Simply place an object, such as a street lamp, between you and the sun and scan the surrounding area. A number of comets have been discovered this way. In 1910 a group of South African diamond miners found a bright comet that had kept itself under wraps by approaching the sun from the opposite direction from us. When

Comet Austin. (Photograph by Jack Newton, August 23, 1982, using a 16-inch f/5 reflecting telescope; 14-minute exposure using Tri-X film.)

the comet rounded the sun, it was discovered quickly by anyone who happened to look toward it. Such finds, it should be pointed out, are extremely rare and also obviously competitive, for a comet bright enough to be seen in daylight will undoubtedly be discovered independently by other observers. But by searching in the sun's vicinity, you may find such a comet first.

It is also important to ensure that your object is a comet, and not a cirrus cloud that is close to the sun. A Montreal amateur astronomer reported a streak of cirrus as a comet early in 1967. Watch the size, shape, brightness, color, and apparent motion of anything you suspect to be a comet.

Horizon at Twilight

Searching for bright comets that are visible in the darkening sky with either the unaided eye or binoculars can be rewarding. With the unaided eye, scan the horizon about a half-hour after the sun has gone down. With binoculars, sweep the western horizon in the evening, or the eastern in the morning, from the horizon to an altitude of about 15 degrees. Sweep slowly with deliberate hand movements every 2 seconds from field to field. Although comets can appear at any time and any location, an evening search is best done just after the full moon has left the area, so the 2–5 nights just after full moon are the ones most likely to bring success. The nights around a new moon are the most promising for hunts in the dawn sky.

Plot any suspects on a sketch of the star field as you see it in your telescope's eyepiece. Note the time. Make sure that the object is not a background deep-sky object plotted on a star atlas or listed in a catalog, or the predicted return of a known comet.

Areas to Skirt

If you use a small telescope, avoid the star-rich areas of the Milky Way, and definitely overlook areas of Coma Berenices, Leo, and Virgo. Otherwise you'll be eyeball-deep in galaxies to check, some of which mimic faint comets. Also the Milky Way contains so many stars that small concentrations of them may look like comets through a small telescope.

Checking Out Suspects

The larger your telescope, the more comet suspects you find, and the more time you will need to spend checking out suspicious objects. Of the two ways of doing this, the most common is simply to locate the object you see in some atlas, like Wil Tirion's *Sky Atlas 2000.0* or his *Uranometria 2000.0*.[2]

If your atlas or catalog shows no fuzzy object where there is clearly one in the field of your telescope, sketch your find's position and continue sweeping, going back in half an hour to check if the object has moved.

Discovering a Comet

If you hunt long enough, sooner or later you will find something that doesn't appear in an atlas. The first thing to remember is to stay calm, for the next minutes will be busy and critical ones. Follow these steps:

1. Sketch the object and its surrounding stars three times: As you see them through you telescope's finder, its low-power eyepiece, and its high-power eyepiece.

2. See if the object has a tail. Although most faint comets do not show obvious tails, there are very few objects in the sky that are not comets but do have tails.

3. Verify that it is not a ghost image of a nearby bright star. Try moving your telescope a bit. If you have a ghost image, it should change its position relative to nearby stars as you move the telescope.

4. Determine that the object is neither a faint star nor group of faint stars. Use high power to rule out this possibility.

5. Check the position in a star atlas or catalog.

6. Note if the object has moved when you look a half-hour later. If there has been no motion by the dawn or the suspicious object's setting has occurred, I strongly advise against sending a telegram. All comets do show motion eventually. You should wait to confirm motion even if you must wait another 24 hours.

7. Locate known comets in the sky. Check even the positions of comets that should be too faint for your telescope, although if you should be the first to notice an unexpected increase in a known comet's brightness, you definitely should report it. Positions of brighter comets are published in the major astronomy magazines, while IAU *Circulars*, the *Minor Planet Circulars*, and the International Comet Quarterly's *Comet Handbook* are good sources for known comets. If you subscribe to the computer service of CBAT, you can ask their program to identify known comets or asteroids in the area around your suspect.

8. If possible have your sighting confirmed by an experienced comet observer of good reputation. Tell that person the object's position, suspected nature, and direction of motion.

9. Once you detect motion, rule out previously known objects in the area and have your observation confirmed locally, it is time to notify CBAT in Cambridge, Massachusetts. Its addresses are:

Telex: TWX 710-320-6842 ASTROGRAM CAM (Use this number for telegrams. If you address your wire to the Smithsonian Astrophysical Observatory, it could be delayed.)

Email: MARSDEN@CFA or GREEN@CFA (.SPAN, .BITNET or .HARVARD.EDU) or EASYLINK 62794505

Marsden and Daniel Green need the following information:

Suspected nature of the object.

Right ascension and declination; although CBAT uses epoch 2000 coordinates, remember to state which epoch you are using.

Direction and rate of motion, preferably this is supplied with a second position taken a half-hour or more after the first. This motion is *very* important, since it may be some time before a clear sky somewhere permits another observer to confirm your finding. By the same token, fuzzy objects on single photographs should be ignored unless you are willing to take a second photograph or search visually to confirm the object's identity and motion.

The comet's magnitude.

Whether the discovery was a visual or a photographic one; if you observed the object both ways, clearly distinguish how you made each observation.

Your description of the object's appearance, including remarks about its size, angular diameter, and shape and length of the tail if you see one.

The date and time of your observation converted to Universal Time or at least with time zone included.

The instrument you used; include the telescope's aperture, type, and magnification. If you are reporting a photographic or CCD find, add the film emulsion (or for electronic observations, the CCD type), exposure time, limiting stellar magnitude of the photo, and the size of the field in degrees or minutes of arc.

Your name as discoverer, your address, and your telephone number.

Remember: Do not send a telegram unless you are certain that the object is a new comet. The philosophy of sending your announcement telegram, as some people have done, just to make sure you are the first, is hardly scientific and may embarrass you. Imagine the chagrin of the observer I know when he reported having "discovered" the Andromeda galaxy. According to Marsden approximately 98 percent of comet discovery reports from unknown observers turn out to be false alarms.[3] Just because an object is fuzzy does not mean that it is a comet, and even if it is a comet, it may be a known one. Galaxies, nebulae, and ghost images of bright stars are often reported as new comets. Again don't bother reporting fuzzy spots on single photographs without some other observer's confirmation.

If your discovery is confirmed, then the comet will be named for you unless others have also found it. Traditionally a comet is allowed a maximum of three names, which represent the first three people to have seen and reported it until the comet's existence is confirmed (see below). Technically you are in competition with other amateurs, professionals using photographic plates and CCD imaging systems, and even orbiting satellites.

HOW COMETS ARE NAMED

A comet is known by as many as three different designations—after its discoverer or discoverers, in the order of its discovery in a certain year and its passage around the Sun.

Discoverer

A comet is named after its discoverer—up to three independent observers or two working together. Prolific comet hunters like Bradfield should have their comets referred to by name and designation to avoid confusion, such as Comet Bradfield 1992b and Comet Bradfield 1992i. The discoverer need not be human: Periodic

Comet Hartley–IRAS 1983v, was first discovered by astronomer Malcolm Hartley and independently by IRAS. A comet orbiting the sun in fewer than 200 years is called a periodic comet. We write Periodic Comet Levy or simply P/Levy. A handful of comets are named not for their discoverers but for those who computed their orbits. The most famous of these is P/Halley, which has been well observed since at least 240 B.C.

Astronomers who discover more than one periodic comet have an Arabic numeral included in the comet's name to indicate which comet is being referred to. Periodic Comet Shoemaker–Levy 7, for example, is the seventh discovered by the team of Carolyn and Eugene Shoemaker and me. Incidentally all of the Shoemaker–Levy comets were discovered by Carolyn on films taken by the team of Eugene and Carolyn Shoemaker and David Levy. Thus it is possible to get your name on a comet if you are part of the discovery team even though you did not actually find the comet.

On rare occasions a comet is discovered by so many observers that no one receives priority as discoverer. Such a comet is usually very bright, and it is generally known by some other name. Examples include the Great Comet of 1843, and the Eclipse Comet of 1948.

Year of Discovery

Immediately on its acceptance as a new or recovered periodic comet by the CBAT, a comet is assigned a designation that includes the year followed by a lower case letter of the alphabet indicating the order of its acceptance by CBAT that year, Comet Levy 1990c, for example. In those years when the number of comets exceeds 26, the alphabet is started again with the subscript one added for each designation, like $1989c_1$. Periodic comets that have a known period of orbital revolution less than 200 years are known by the name or names of their discoverers, preceded by a P/. Some periodic comets that are observed throughout their whole orbits do not receive a year-and-letter designation, including P/Encke,

P/Gunn, P/Arend–Rigaux, and P/Schwassmann–Wachmann 1. Such comets are sometimes called annual comets.

If a discoverer (like Bradfield) or a discovery team (like Shoemaker–Levy) finds more than one periodic comet, a number follows the comet. Periodic Comet Shoemaker–Levy does not exist, for instance. Following the rules, the comet became Periodic Comet Shoemaker–Levy 1 when the second periodic comet was found. Now there is a Periodic Comet Shoemaker–Levy 9. There is however only one Periodic Comet Levy.

Occasionally a periodic comet that had been considered lost (since it wasn't seen for several predicted apparitions) is accidentally rediscovered and then identified as the missing comet. It may then be assigned the rediscoverer's name in addition to the first discoverer's name, as happened recently with 1991a, P/Metcalf–Brewington. Previously known as Periodic Comet Metcalf, it had not been seen since its discovery by J. H. Metcalf in 1906. Howard Brewington, an amateur astronomer hunting from New Mexico, rediscovered it in 1991.

Passage Around the Sun

Some time after the end of each year, all comets reaching perihelion in that year receive a Roman numeral designation giving the order in which they reached perihelion. Thus, P/Halley in 1910 was the second comet to reach perihelion in 1910, and is known as 1910 II. P/Halley is also known as 1759 I and 1835 III, as well as 1982i. Now following its 1986 perihelion, it is also referred to as 1986 III.

ELEMENTS OF AN ORBIT

Every planet, comet, asteroid, and tiny meteoroid in the solar system has a path we call an orbit. We know the orbits of the nine major planets very well, since we have observed them over many years. As each new comet is found, the first thing we want to know is its orbit, which is really a definition of the comet's place in space over a period of time.

An equinox is that point in the Earth's orbit when the Earth's equatorial plane meets the orbital plane, a plane we call the ecliptic. (The ecliptic is the apparent path of the center of the sun around the sky.) The Earth crosses that point twice each year, once in March and once in September. On about March 21 of each year, the Earth crosses the equinox as the sun moves north. In the northern hemisphere, we call this point of the Earth's orbit the vernal equinox. We use the vernal equinox as an important reference point.

We picture the orbits of comets in relation to the orbit of Earth. With at least three (and preferably a lot more) precise positions of the comet, we can determine the six orbital elements we need to describe a comet's journey through the solar system. Admittedly this is a somewhat oversimplified explanation of a complex process. Here are the six basic orbital elements:

The comet's perihelion date T: The date and time the comet is closest to the sun.

How close the comet will be to the Sun at perihelion: This factor, q, is expressed in terms of a yardstick called an astronomical unit—approximately the average distance between Earth and sun.

The eccentricity e: If the orbit were a perfect circle, e would be 0; if the orbit were a parabola, e would be 1. The earliest orbits for most comets are figured as if $e = 1$. If the comet turns out to be periodic, traveling in an ellipse, it will quickly deviate from the predicted path in the days and weeks after discovery.

How far the comet is inclined to the ecliptic: This angle of inclination is called i. Most periodic comets have inclinations less than 20 degrees. Long-period comets can have any inclination. Inclinations of 90–180 degrees refer to comets moving backward with respect to those with inclinations from 90–0 degrees.

The longitude of the ascending node (Ω): This is the angular distance between the vernal equinox and the point, which we call the ascending node, where the comet crosses the plane of the Earth's orbit as the comet moves northward.

The argument of perihelion (ω): This is the separation, measured again as an angle, between the ascending node and the point where the comet is closest to the sun. Finally all these elements are accurate only at one time, which we call the epoch. Orbits can be computed for other epochs of interest to observers.

For example, here is a set of orbital elements for Comet Shoemaker–Levy 1993h, as described in *IAU Circular* 5808.[4] My comments are in brackets.

COMET SHOEMAKER–LEVY (1993h)

[The circular begins with a description of the observations, followed by accurate positions of the comet.]

The following precise positions have been measured by B. A. Skiff and C. S. Shoemaker from exposures by C. S. Shoemaker, E. M. Shoemaker and D. H. Levy with the 0.46-m Schmidt telescope at Palomar (the first R.A. [right ascension; the position in the sky roughly equivalent to longitude on Earth] and the second Decl. [declination; the position in the sky roughly equivalent to latitude on Earth] are uncertain, and the third image was involved with a star): [It was hard to measure the comet, since a star was almost obscuring it.]

1993 UT R.A. (2000) Decl. [Because the Earth precesses, the positions need to be given relative to what the coordinate system is at a certain time or equinox; this one is the year 2000.]

Circular No. 5808

Central Bureau for Astronomical Telegrams
INTERNATIONAL ASTRONOMICAL UNION

Postal Address: Central Bureau for Astronomical Telegrams
Smithsonian Astrophysical Observatory, Cambridge, MA 02138, U.S.A.

Telephone 617-495-7244/7440/7444 (for emergency use only)
TWX 710-320-6842 ASTROGRAM CAM   EASYLINK 62794505
MARSDEN@CFA or GREEN@CFA (.SPAN, .BITNET or .HARVARD.EDU)

COMET SHOEMAKER-LEVY (1993h)

The following precise positions have been measured by B. A. Skiff and
C. S. Shoemaker from exposures by C. S. Shoemaker, E. M. Shoemaker and
D. H. Levy with the 0.46-m Schmidt telescope at Palomar (the first α and
the second δ are uncertain, and the third image was involved with a star):

1993	UT	α_{2000}	δ_{2000}
May	23.20885	13ʰ24ᵐ50ˢ.49	−33°58'19".7
	23.24809	13 24 48.61	−33 58 22.2
	24.20572	13 24 01.91	−33 59 22.2
	24.23836	13 24 00.19	−33 59 23.0
	26.20329	13 22 26.46	−34 01 18.4
	26.25729	13 22 23.84	−34 01 21.2

Preliminary parabolic orbital elements by B. G. Marsden, Harvard-
Smithsonian Center for Astrophysics:

$$T = 1993 \text{ May } 14.144 \text{ TT}$$
$$\omega = 199.384$$
$$\Omega = 31.929 \Big\} 2000.0$$
$$q = 5.41446 \text{ AU}$$
$$i = 72.368$$

1993 TT		α_{2000}	δ_{2000}	Δ	r	ϵ	β	m_1
May	13	13ʰ33ᵐ.84	−33°44'.4	4.503	5.414	151.8	5.1	16.6
	23	13 25.01	−33 58.1	4.568	5.415	143.4	6.4	16.6
June	2	13 17.45	−34 06.8	4.660	5.416	134.2	7.7	16.7
	12	13 11.38	−34 13.5	4.773	5.419	124.9	8.8	16.7
	22	13 06.90	−34 20.7	4.904	5.422	115.7	9.7	16.8
July	2	13 04.04	−34 30.7	5.047	5.426	106.6	10.3	16.9
	12	13 02.71	−34 45.2	5.198	5.432	97.9	10.7	16.9
	22	13 02.83	−35 05.4	5.352	5.438	89.5	10.8	17.0
Aug.	1	13 04.25	−35 32.1	5.504	5.446	81.4	10.6	17.1

NOVA AQUILAE 1993

Visual magnitude estimates: May 21.04 UT, 8.0 (T. Vanmunster, Lan-
den, Belgium); 21.97, 7.8 (H. Dahle, Oslo, Norway); 23.96, 7.5 (S. Baroni,
Milan, Italy); 25.97, 7.9 (B. H. Granslo, Fjellhamar, Norway); 26.95, 8.0
(Dahle); 28.96, 8.3 (J. D. Shanklin, Cambridge, England).

1993 May 31   *Daniel W. E. Green*

May	23.20885	13 24 50.49	−33 58 19.7
	23.24809	13 24 48.61	−33 58 22.2
	24.20572	13 24 01.91	−33 59 22.2
	24.23836	13 24 00.19	−33 59 23.0
	26.20329	13 22 26.46	−34 01 18.4
	26.25729	13 22 23.84	−34 01 21.2

Preliminary parabolic orbital elements by B. G. Marsden, Harvard–Smithsonian Center for Astrophysics:

T = 1993 May 14.144 TT Argument of Perihelion. = 199.384
[Longitude of Ascending] Node = 31.929 2000.0

q = 5.41446 AU   Incl. = 72.368

[Finally from the elements that follow, the comet's future path is predicted.]

1993 TT	R. A. (2000)	Decl. Positions
May 13	13 33.84	−33 44.4
23	13 25.01	−33 58.1
June 2	13 17.45	−34 06.8
12	13 11.38	−34 13.5
22	13 06.90	−34 20.7
July 2	13 04.04	−34 30.7
12	13 02.71	−34 45.2[4]

COORDINATES IN THE SKY

Celestial coordinates are a projection onto the sky of the co-ordinate system of Earth. The celestial equator is a projection of Earth's equator, and the celestial poles are projections of Earth's rotational poles.

Declination in the sky is measured like earthly latitude, in degrees, minutes, and seconds north and south of the equator. Right ascension is equivalent to longitude, it is measured in units of time—hours, minutes, and seconds, rather than degrees. The zero hour of right ascension is the position of the Sun at the first moment of spring in the northern hemisphere—the *vernal* equinox—and it is along the line of intersection of the Earth's equatorial plane and its orbital plane or ecliptic. Hours in right ascension are measured eastward from the vernal equinox.

Since the Earth's poles wobble because of tidal effects of the moon and sun, the direction of the vernal equinox shifts in the sky. The entire coordinate system shifts with that wobble. The wobble is called precession. Hence we must find a precise moment in time to define our celestial coordinates; that moment is called the epoch.

THE MAGNITUDE SCALE

How do we describe the brightness of stars, planets, and comets? Anyone going outside on a clear night will notice that not all the stars are the same brightness. Our system of magnitudes dates back to the second century B.C. Greek astronomer Hipparchus, who divided the stars into six brightness groups, with the 20 brightest stars called first magnitude and the faintest stars sixth magnitude. By 1856 Norman Pogson of Radcliffe Observatory had quantified this relationship, making a first-magnitude star 100 times brighter than the faintest star visible without a telescope, a second-magnitude star 2.5 times fainter than a first magnitude star, and a third-magnitude star, 2.5 times fainter again.

Vega, the brightest star in the summer triangle, is a zero-magnitude star. Pogson defined Polaris, the North Star, as being second magnitude. Most of the stars in the Big Dipper are also about second magnitude. Most of the stars in nearby Cassiopeia are a magnitude fainter.

Why are the stars of different brightnesses? There are two reasons. Like lights on a street, stars are fainter the further away they are, and some stars are intrinsically more luminous than other stars. The result is a hodgepodge of stars of different brightnesses.

Because comets are not points of light but fuzzy objects that cover an extended area of sky, their magnitudes are not so easy to determine. In trying to estimate the brightness of a comet, observers usually compare the comet to out-of-focus stars of known magnitude. Observers deliberately unfocus their telescopes so that stars as bright as the comet appear the same size as the comet. The procedure takes practice, but when we say that a comet is magnitude 7.1, that is how we arrive at that estimate.

References

CHAPTER 1

1. G. W. Hough, "Observations of Comet II. 1862, made with the Olcott Meridian Circle, at the Dudley Observatory," *Astronomische Nachrichten* **59**, no. 1394 (1863), col. 29.
2. *Ibid.* See also S. K. Vsekhsvyatskii, *Physical Characteristics of Comets* (Jerusalem: Israel Program for Scientific Translations, 1964), 215; *Illustrated London News*, Aug. 16, 1862, 179.
3. H. P. Tuttle, "Schreiben des Herrn Tuttle an den Herausgeber," *Astronomische Nachrichten* **59**, no. 1404 (1863), cols. 187–90.
4. Many of the interesting details about Horace Tuttle's life come from an unpublished study circa 1980 "H. P. Tuttle: Cometseeker," by Richard E. Schmidt, US Naval Observatory.
5. The basis for this story comes from an unpublished memoir (Ref. 4).
6. W. T. Lynn, "Comet III. 1862," the *Observatory* **25** (1902), 304–305. The author notes that the *Astronomische Nachrichten*, the journal of record at the time, called it Comet II. 1862 but that it was the third comet to reach perihelion that year and is correctly III. 1862 (or in today's parlance, 1862 III).
7. B. G. Marsden, "The Next Return of the Comet of the Perseid Meteors," *Astronomical Journal*, **78**, 656 (1973).
8. B. G. Marsden, *Astronomical Journal* **78**, 654–662 (1973); see also *IAU Circulars* 5330 (August 28, 1991) and 5586 (August 14, 1992).
9. *IAU Circular* 5330 (August 28, 1991).
10. B. G. Marsden, personal communication, Oct. 18, 1992.

11. B. G. Marsden, "Comet Swift–Tuttle: Does It Threaten Earth?" *Sky and Telescope* **85**, 1 (1993), 11.
12. *IAU Circular* 5620, Sept. 27, 1992.
13. B. G. Marsden, personal communication, Oct. 18, 1992.
14. B. G. Marsden, *Astronomical Journal* **78**, 658 (1973).
15. B. G. Marsden, personal communication, Oct. 12, 1992. See also *Time Magazine*, Nov. 9, 1992, 27.
16. B. G. Marsden, personal communication, Oct. 31, 1992.
17. B. Marsden, interview, 7 January 1993.

CHAPTER 2

1. *Julius Caesar*, II, ii, 29–30.
2. Chronicles I, 21:16.
3. A. A. Barrett, "Observations of Comets in Greek and Roman Sources before A.D. 410," *Journal of the Royal Astronomical Society of Canada* 72 (1978), 81–106.
4. Seneca, *Quaestiones Naturales*, VII, "De Cometis" XVII, 2.
5. Seneca, *Quaestiones Naturales* VII, "De Cometis" I, 1.
6. Seneca, I, 4–5.
7. Seneca, *De Cometis* XVI, 2.
8. R. F. Rogers, "Newly Discovered Byzantine Records of Comets," *Journal of the Royal Astronomical Society of Canada* **56**, 5 (1952), 177.
9. Seneca, *De Cometis* XXI, 2–3.
10. *Ibid.*, XXIII, 1–3.
11. *Ibid.*, XVII, 1.
12. The *Works of Tacitus*, Oxford translation (London: G. Bell and Sons, 1910), 367–68.
13. *The Works of Tacitus.*[11]
14. *Illustrated London News*, July 1861, 65.

CHAPTER 3

1. *The Poetical Works of Gerard Manley Hopkins*, ed. N. H. MacKenzie (Oxford: Clarendon Press, 1990), 40.
2. J. R. Hind, Letter, *Times*, Aug. 1, 1864, 5, col. e.
3. S. Drake, *The Controversy on the Comets of 1618* (Philadelphia: University of Pennsylvania Press, 1960), 7.
4. S. Drake, 25.
5. S. Drake, 19. The thought from Horace is from *Carmina* I, i, 36.

6. J. Swan, *Speculum Mundi: or a Glasse Representing the Face of the World* (Cambridge, England: by the Printers to the University, 1635), 80–81.
7. L. C. Peltier, *Starlight Nights: The Adventures of a Star Gazer* (Cambridge, MA: Sky Publishing Corp., 1980), 17.
8. *Ibid.*
9. B. G. Marsden, personal communication, Oct. 1, 1992.
10. J. R. Hind, The Comet of 1556; Being Popular Replies to Everyday Questions, Referring to Its Anticipated Reappearance, with Some Observations on the Apprehension of Danger from Comets (London: John W. Parker and Son, 1857), 1.
11. *The London Times,* July 4, 1861, p. 12.
12. The *London Review,* July 13, 1861, 46, col. 2.
13. Hind, 14.
14. *Ibid.,* 45.
15. B. G. Marsden, personal communication, Oct. 1, 1992.

CHAPTER 4

1. C. A. Lubbock, *The Herschel Chronicle* (Cambridge, England: Cambridge University Press, 1933), 15.
2. *Ibid.,* 15.
3. *Ibid.,* 60.
4. *Ibid.,* 66.
5. *Ibid.,* 66.
6. W. Herschel announced his discovery at the end of March 1781 to the Bath Literary and Philosophical Society. This "Account of a Comet" appears in *The Scientific Papers of Sir William Herschel*, J. L. E. Dreyer, ed. (London: The Royal Society and the Royal Astronomical Society, 1912), 1: 30–38.
7. W. G. Hoyt, *Planets X and Pluto* (Tucson: University of Arizona Press, 1980), p. 12.
8. Charles Messier to William Herschel, in Lubbock, 86.
9. Lubbock, p. 95.
10. W. Herschel to C. Herschel, July 3, 1782. See also "America's Last King and His Observatory" in J. Ashbrook, *The Astronomical Scrapbook* (Cambridge, MA.: Sky Publishing Corp., 1984), 17.
11. Lubbock, 246.
12. N. A. Mackenzie, "He Broke through the Barriers of the Skies," *Sky and Telescope* 8, no. 5 (1949), 119. The title is a translation of the words on Herschel's tombstone, Coelorum perrupit claustra. *See also;* Lubbock, p. 177.
13. P. M. Millman, "The Herschel Dynasty" *Journal of the Royal Astronomical Society of Canada* 74 (1980), 211.
14. Hoyt, 15.

CHAPTER 5

1. K. G. Jones, *Messier's Nebulae and Star Clusters*, 2d ed. (London: Cambridge University Press, 1991), 347.
2. Although this story appears in several sources, a particularly good one is K. G. Jones, 365. Some of the details of Messier's relationship with de Saron come from this source as well.
3. If we count some independent discoveries of comets where he was not the first finder, several others for which he was far too late to have his name attached, two discoveries years apart of the periodic comet now named after Encke, and comets that were never confirmed by other observers, his total is about 37.
4. R. K. Marshall, "Astronomical Anecdotes," *Sky and Telescope* 3 (Apr. 1944), 19.
5. John Keats, "On First Looking into Chapman's Homer".
6. J. Ashbrook, "Harvester of the Skies," *Sky and Telescope* 25 (1963), 198–199.
7. A. Ewing, *Practical Astronomy* (Burlington, NJ: David Allinson, 1812).
8. Different parts of this quotation appear in two excellent Mary Proctor books on comets: *The Romance of Comets* (New York: Harper and Brothers, 1926), 27; M. Proctor and A. C. D. Crommelin, *Comets: Their Nature, Origin, and Place in the Science of Astronomy* (London: Technical Press, 1937), 154–55.
9. See J. Bortle, "Comet Digest," *Sky and Telescope* 64 (1982), 294.
10. At Tennessee's Arthur J. Dyer Observatory, Robert Hardie noted in "The Story of the Early Life of E. E. Barnard" that the university named one of its dormitories after the great comet finder, who was known not to need much sleep.
11. H. B. Curtis, "The Comet-Seeker Hoax," *Popular Astronomy* 46 (1938), 70.
12. *Ibid.*, 70.
13. *Ibid.*
14. *Ibid.*, 75.
15. *Ibid.*, 75.
16. *Ibid.*
17. J. Lankford, "E. E. Barnard and the Comet-Seeker Hoax of 1891," *Sky and Telescope* 57 (1979), 420–22.
18. Curtis, 71.

CHAPTER 6

1. H. D. Thoreau, *Walden*, 1854; W. Harding, ed. *The Variorum Walden* (New York: Twayne, 1962), 259, 261.
2. Jones, 371.
3. *Ibid.*, Peltier, 237.
4. Proctor and Crommelin, 112.

5. *Ibid.*, 117–18.

6. *Observatory* 4, 331 (1881).

7. This story comes from G. Kronk, *Comets: A Descriptive Catalog* (Hillside, NJ: Enslow, 1984), 120.

8. Peltier, 36.

9. *Ibid.*, 127.

10. *Ibid.*, 134.

11. F. L. Whipple, *The Mystery of Comets* (Washington, DC: Smithsonian Institution Press, 1985), 145–147.

12. Whipple announced his theory in two papers. "A Comet Model. I. The Acceleration of Comet Encke," *Astrophysical Journal* 111 (1950), 375–94, explains how the orbit of periodic Comet Encke, which is shrinking with each return, is interpreted if the structure of its nucleus consists of meteoric material embedded in ices that sublimate to gases. As the freed material rushes out of the comet with some force, it can accelerate the comet. The second paper, "Physical Relations for Comets and Meteors" *Astrophysical Journal* 113 (1951), 464–74, expands on this model.

13. R. Gore, "Much More Than Met the Eye: Halley's Comet '86," *National Geographic* 170 (1986), 778.

14. *IAU Circular* 1924, Sept. 28, 1965.

15. *Circular* 1925, Oct. 1, 1965. See also B. G. Marsden, personal communication, Jan. 2, 1993.

16. B. G. Marsden, "The Sun-Grazing Comet Group," *Astronomical Journal* 72 (1967) 1179.

17. *Circular 1932*, Oct. 26, 1965, and personal communication, August 1993

18. P. L. Brown, *Comets, Meteorites, and Men* (London: Robert Hale and Co., 1973), 91.

19. *Circulars 1937*, Nov. 9, 1965, and 1949, Feb. 8, 1966. The suggested periods were 830 and 1110 years.

20. W. Houston, personal communication.

21. Peltier, 231.

CHAPTER 7

1. B. G. Marsden, "The Sun-Grazing Comet Group," *Astronomical Journal* 72 (1967), 1170–83.

2. See B. G. Marsden, "Catalogue of Discoveries and Identifications of Minor Planets" (rev. Identifizierungsnachweis der Kleine Planeten) 2d ed. (Minor Planet Center, International Astronomical Union, 1986).

3. Marsden to Alcock, April 13, 1983.

CHAPTER 8

1. D. Levy, "The Art of Comet Hunting," *Royal Astronomical Society of Canada*, **65** (1970), L8.
2. C. Messier, "A Catalogue of Nebulae and Star Clusters Discovered among the Fixed Stars above the Horizon of Paris. Observed at the Observatory of the Navy with different instruments by M. Messier." *Historire de l'Academie Royale des Sciences* (1771). Numbers 1–45 appeared here; Numbers 46–68 appeared in *Connaissance des Temps*, (1781); Numbers 69–103 surfaced in *Connaissance des Temps*, (1784); 104–109, added to by Méchain, in *Berliner Astronomisches Jahrbuch*, (1786). Messier 110, a companion to M31, the Andromeda Galaxy, was added a few years ago, based on solid evidence that Messier had drawn it.
3. D. H. Levy, *Clyde Tombaugh: Discoverer of Planet Pluto* (Tucson: University of Arizona Press, 1991).
4. C. Tombaugh, plate envelope notes for No. 171, courtesy Lowell Observatory.
5. One of the unsung heroes of asteroidal astronomy, Conrad Bardwell has worked at the Minor Planet Center since 1958. He has contributed to the computerization of the Minor Planet Center and was among the first to use a computer to find minor planets. B. G. Marsden, personal communication, Jan. 22, 1993.
6. B. G. Marsden, "Original and Future Cometary Orbits IV," *Astronomical Journal* **99** (1990), 1971–73. As the original comet moved away from the Sun, the nongravitational force of the gases leaving the comet decreased, but it was still sufficient to allow Comets 1988e and 1988g to separate.

CHAPTER 9

1. G. K. Gilbert, "The Origin of Hypotheses: Illustrated by the Discussion of a Topographic Problem," presidential address, the Geological Society of Washington, March 1896. See also *Science*, N. S. 3 (1896), 1.
2. *Ibid.*, 11.
3. E. M. Shoemaker, interview, Oct. 23, 1992.
4. A version of this story first appeared in *Sky and Telescope* **83** (1992), 219–21.

CHAPTER 10

1. E. M. Shoemaker, interview, Jan. 27, 1993.
2. E. M. Shoemaker, interview, Jan. 27, 1993.

CHAPTER 11

1. Stubbs, ed., *Gervasii Cantuariensis Opera Historica: Chronica Gervasii, Rerum Britannicarum Medii Aevi Scriptores*, (London, 1879), 73a.
2. J. B. Hartung, "Was the Formation of a 20-km Diameter Crater on the Moon Observed on June 18, 1178?" *Meteoritics* 11 3, (1976), 187–94. Calame and Mulholland, cited below, state that the correct Gregorian date is June 25, 1178.
3. K. Brecher, "The Canterbury Swarm: Ancient and Modern Observations of a New Feature of the Solar System," *Bulletin of the American Astronomical Society* 16 (1984), 476.
4. B. E. Schaefer, "The 'Lunar Event' of A.D. 1178: A Canterbury Tale?" *Journal of the British Astronomical Association* 100 (1990), 211.
5. W. G. Waddington, "More on the 'Canterbury Event' of 1178," *Journal of the British Astronomical Association* 101 (1991), 79.
6. E. M. Shoemaker, interview, Feb. 11, 1993.
7. E. M. Shoemaker.[6]
8. R. L. Heacock, G. P. Kuiper, E. M. Shoemaker, H. C. Urey, and E. A. Whitaker, "Technical Report No. 32-800: Ranger VIII and IX" (Pasadena: Jet Propulsion Laboratory, 1966), 2.
9. Apollo 11 and NASA; July 20, 1969.
10. E. M. Shoemaker, interview, Feb. 17, 1993.
11. E. M. Shoemaker.[10]
12. E. M. Shoemaker.[10]
13. O. Calame and J. D. Mulholland, "Lunar Crater Giordano Bruno: A. D. 1178 Impact Observations Consistent with Laser Ranging Results," *Science* 199 (1978), 875.
14. F. K. Duennebier, Y. Nakamura, G. V. Latham, and H. J. Dorman, "Meteoroid Storms Detected on the Moon," *Science* 192 (1976), 1000–02.

CHAPTER 12

1. E. F. Helin and E. M. Shoemaker, "The Palomar Planet-Crossing Asteroid Survey, 1973–1978," *Icarus* 40 (1979), 321–28.
2. L. D. Schmadel, *Dictionary of Minor Planet Names* (Berlin: Springer-Verlag, 1992), 593; *cf.* MPC 19338.
3. D. Morrison and M. Shapley Matthews, *Satellites of Jupiter* (Tucson: University of Arizona Press, 1981), 435, 438–451.

CHAPTER 13

1. H. Wright, *Palomar: The World's Largest Telescope* (New York: MacMillan, 1952), 135.

2. C. S. Shoemaker, interview, Nov. 30, 1992.
3. C. S. Shoemaker.[2]
4. *Sky and Telescope* 81 (1991), 659.

CHAPTER 14

1. J. M. Greenberg, "What are Comets Made of? A Model Based on Interstellar Dust," in *Comets*, ed. L. Wilkening (Tucson: University of Arizona Press, 1982), 131, 157.
2. G. A. Soffen, "Life in the New Solar System?" in *The New Solar System*, 3d ed., eds. K. Beatty and A. Chaikin (Cambridge, MA: Sky Publishing Corp., 1990), 276.
3. J. N. Marcus and M. A. Olsen, "Biological Implications of Organic Compounds in Comets," in *Comets in the Post-Halley Era*, 449.
4. "Comets and Meteorites: Harbingers of Life on Earth," *Sky and Telescope* 78 (1989), 242.
5. Benton C. Clark, "Primeval Procreative Comet Pond," *Origins of Life and Evolution of the Biosphere* 18, (1988), 209–38.
6. V. A. Firsoff, *Life among the Stars* (London: Allan Wingate, 1974), 31.
7. T. R. Cech, "RNA as an enzyme," *Scientific American* 255 (1986), 64–75.
8. M. Bailey, S. Clube, and W. Napier, *The Origin of Comets* (Oxford: Pergamon, 1990), 454–55.

CHAPTER 15

1. D. A. Russell, "The Disappearance of the Dinosaurs," *Canadian Geographical Journal* 83, 6 (1971), 204–15.
2. D. A. Russell, "The Environments of Canadian Dinosaurs," *Canadian Geographical Journal* 87 (1973), 4–11.
3. L. Alvarez, W. Alvarez, F. Asaro, and H. Michel, "Extraterrestrial Cause for the Cretaceous–Tertiary Extinction," *Science* 208 (1980) 1095–1107.
4. Alvarez *et al.*, 1102–1104.
5. C. B. Officer and C. L. Drake, "Terminal Cretaceous Environment Events," *Science* 227, 1161–67.
6. Officer and Drake, 1165.
7. V. L. Sharpton, G. P. Dairymple, L. E. Marin, G. Ryder, B. C. Schuraytz, and J. Urrutia-Fucugauchi, "New links between the Chicxulub Impact Structure and the Cretaceous–Tertiary Boundary," *Nature* 359, 6398 (1992), 819–21.
8. "Possible Yucatan Impact Basin," *Sky and Telescope* 63 (1982), 249–50.

9. The story about the Yucatan Crater includes details from J. Kelly Beatty's "Killer Crater in the Yucatan?" *Sky and Telescope* **82** (1991), 38–40.

CHAPTER 16

1. D. di Cicco, "New York's Cosmic Car Conker," *Sky and Telescope* **85**, 2 (1993), 26.
2. J. K. Davies, *Cosmic Impact* (London: Fourth Estate, 1986), 185.
3. *Ibid.* 169.
4. Z. Ceplecha, "Earth-Grazing Daylight Fireball of Aug. 10, 1972," Hazards Due to Comets and Asteroids Conference, Tucson, AZ, Jan. 5, 1993.
5. E. M. Shoemaker, "Asteroid and Comet Bombardment of the Earth," *Annual Review of Earth and Planetary Sciences* **11** (1983), 476.
6. Shoemaker, 480.
7. "The Tunguska Meteorite and Atmospheric Ozone," *Sky and Telescope* **63** (1982), 14.
8. C. R. Chapman, "Hazard to Civilization of Asteroid and Cometary Impacts," Asteroid Hazard Conference, USSR Academy of Sciences, St. Petersburg, Oct. 10, 1991.
9. D. Morrison and C. R. Chapman, "The Nature of the Impact Hazard," Hazards due to Comets and Asteroids Conference, Tucson, AZ, Jan. 5, 1993.
10. R. A. F. Grieve and E. M. Shoemaker, "The Record of Past Impacts on Earth," Hazards Due to Comets and Asteroids Conference, Tucson, AZ, Jan. 5, 1993.
11. E. M. Shoemaker and C. S. Shoemaker, "Collision Rate of Comets with Earth," Hazards Due to Comets and Asteroids Conference, Tucson, AZ, Jan. 5, 1993.
12. B. G. Marsden, personal communication, May 11, 1993.
13. C. Chapman, "Hazard to Civilization of Asteroid and Cometary Impacts," Asteroid Hazard Conference USSR Academy of Sciences, St. Petersburg, Oct. 10, 1991.

CHAPTER 17

1. F. Hoyle, *The Black Cloud* (New York: Signet, 1959), 16. Two-and-a-half degrees are roughly equivalent to five full-moon diameters.
2. F. Hoyle.[1]
3. T. Gehrels, *On the Glassy Sea: An Astronomer's Journey* (New York: American Institute of Physics, 1988), 108, and personal communication.
4. D. Morrison, lecture at Hazards Due to Comets and Asteroids Conference, Tucson, AZ, Jan. 5, 1993.

5. San Jose *Mercury News,* Mar. 22, 1992.
6. P. R. Weissman, "Scientific Objectivity and the Impact Hazard: Responsible Reporting vs. Crying Wolf," Hazards Due to Comets and Asteroids Conference, Tucson, AZ, Jan. 5, 1993.
7. D. Lindley, "Earth Saved from Disaster!" *Nature* **360** (1992), 623.
8. The *New York Times,* Nov. 3, 1992, B5.

CHAPTER 18

1. J. Scotti to B. G. Marsden, March 26, 1993.
2. C. Chapman, "Comet on Target for Jupiter," *Nature* **363** (1993), 492.
3. J. Scotti, personal communication, Aug. 29, 1993.
4. *IAU Circular* 5801, May 22, 1993.

EPILOGUE

1. A. P. Hankins, *Cole of Spyglass Mountain* (New York: Dodd, Mead, and Co., 1923).

AFTERWORD

1. R. Burnham to D. Levy, June 7, 1967.
2. W. Tirion, *Sky Atlas 2000.0* (Cambridge, MA: Sky Publishing Corp. and Cambridge, England: Cambridge University Press, 1981). W. Tirion, B. Rappaport, and G. Lovi, *Uranometria 2000.0,* vols. I and II (Richmond, VA: Willmann-Bell, 1987–1988).
3. B. G. Marsden, personal communication, July 17, 1993.
4. *IAU Circular* 5808, May 31, 1993.

Index